I dedicate this book to our Baby Saint, whom we trust is in heaven following a miscarriage. Baby Saint didn't have the opportunity to live the "good life" on earth with his Mom and Dad and brothers and sister in the way that we experience, nor did we receive the gifts Baby Saint would have provided us but we do believe that Baby Saint is living the ultimate "good life" in heaven—for which we all strive. Surely Baby Saint's intercession from heaven has helped us along our way.

Bert Tranel

River Lights Publishing
1098 Main St.
Dubuque, IA 52001

Copyright © Bert Tranel

All rights reserved including the right of reproduction in whole or in part in any form. No part of this publication may be reproduced or transmitted in any form or by any means without written permission of the author.

Book design by Alina Crow Designs

Manufactured in the United States of America

ISBN: 978-1-7342780-2-6

Printed in the United States of America

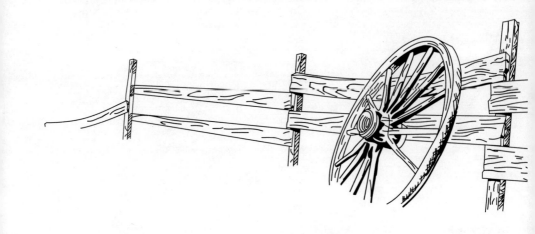

Primitive to Modern Farming

Living the Good Life

Bert Tranel

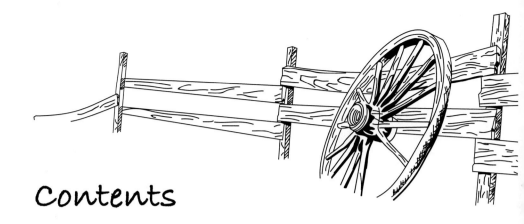

Contents

Introduction   vii

Chapter 1
Hunting and Gathering   1

Chapter 2
Evolution of Growing Grain   7
　Part 1: Wheat   7
　Part 2: Corn   13
　Part 3: A Brief History of Corn Planters   17

Chapter 3
Agricultural Industrial Revolution (1760–1850)   23

Chapter 4
Some Family History of Farmers   27

Chapter 5
Some Local Flavor   31

Chapter 6
Beginning Grampo's Era   37

Chapter 7
Farm Buildings   43

Chapter 8
Marketing Livestock   53

Chapter 9
Livestock Sickness and Treatments   61

Chapter 10
Farm Sounds of The Past   69

Chapter 11
A Final Sale   75

Chapter 12
Future Farmers   79
   Part 1: Milking Robots   79
   Part 2: Cattle Ranching   83
   Part 3: Grain Production   87
   Part 4: A Midwest Mega Grain Farmer   89
   Part 5: Hog Production in 2019   92
   Part 6: Poultry Production   94
   Part 7: A Word About Sheep   98

Chapter 13
Weed Control   101

Epilogue   *107*

Introduction

When Ann, my wife, and I get together with her siblings and mine and other relatives, many of whom are 70 plus years of age, our conversations most often lead to reminiscing about our younger days on the farm, especially since the majority of us were raised on farms. The "Good Old Days" seems to be the impression left on us, regardless of how good or bad they were. Somehow, we remember the good over the bad. The hard work, the lack of conveniences, etc., that we enjoy today were part of our accepted lifestyle then and now that we made it through those years, we see them through rose-colored glasses. Our conversations soon move on to the many agricultural changes in our lifetimes. I would note that these agricultural changes are likely the prime subject among all older farmers.

Today changes are so fast that they remind me of riding on a fast-moving Amtrak train at 90 miles per hour. You have to take a quick look as the scenery will soon change. This reminds me of an incident I had while riding an Amtrak through North Dakota. I was sitting on the aisle side with a tall, thin gentleman with a grey goatee sitting to my right along the window. I could see with one eye that he was either a rancher or a cowboy with his well-worn boots and a ten-gallon hat. He was wearing Levi jeans, a plaid shirt with snap buttons, and a shiny belt buckle, one like rodeo cowboys wear after winning an event. We struck up a conversation that after a couple hours ran dry (about as dry as the North Dakota prairie we were riding through) at which time he looked out the window and started calling out numbers such

vii

as 68, 32, 48, 36, etc. I asked him why he was calling out numbers. He said he was counting cows in the pastures. I asked how he could count cows when we were going 90 plus miles per hour. His response was, "It's easy, just count their milky things and divide by four."

I have heard it said that writing is like talking on paper. Although much of early history has been handed down by word of mouth, talking about it on paper will keep it from being forgotten. For various reasons, I like to "remember" on paper rather than speaking before an audience. One reason is that if I say the wrong thing, I can delete it and start over. Another is that I don't have to be concerned about getting butterflies in my stomach in front of a live crowd, and finally, I feel it is the best way to keep the past alive. Agricultural change is a subject that could fill many, many thick books. I want to talk on paper about a small segment of changes leading up to, during and after my Grampa Kieffer's farming career. In grampa's day, farming from 1896 to 1930, nearly half of the population was engaged in farming. Today, some statistics estimate about 1% of our population is involved in farming.

Today the young generation, especially those in large cities, knows that food comes from the grocery stores, but ask them where the grocery stores obtain it, you might get a blank look or a highly imaginable response. Although I do need to add that many farmers and farm organizations are working hard to make non-farm people more aware of food production. Our supply of food today is in such abundance that many fortunate people take it all for granted. However, those older folks who lived through the Great Depression and World War II, when food was rationed, have a greater appreciation for our abundance of food.

I am a lover of history, especially the history of agriculture and I enjoy talking about it. I want to preserve some of that heritage. In my lifetime, there has been and still is so much progress being made in the agricultural world, but I don't want to have the dedicated, hard work and truly good aspects of early farming and farmers forgotten. Much of what I am talking about centers around my grampo's time, but I wanted to go back about 10,000 years, the era when all people were hunters and gatherers. What changes will appear during the next 100 years or even the next ten years or as few as five years from now could be a story and a half for another day. It is beyond my wildest imagination, just as the technology of today would have been total science-fiction in grampo's day, a mere 100 years ago. I hope you enjoy "listening" to me and I feel confident you will.

Chapter 1

Hunting and Gathering

Approximately 10,000 years ago, agriculture or farming as we know it today was just getting underway. Before this time, all people were hunters and gatherers. For them to survive, food had to be obtained by catching fish, hunting wild animals, gathering nuts, plants, seeds, fruits, and other edibles depending on what part of the world they were in. For the most part, hunters and gatherers were restricted to eating small game as they lacked weapons to kill large animals. An exception would have been if larger animals died of natural causes or by a predator, or by stampeding animals off a cliff which was a method often used by the American Indians to kill buffalo. Eventually technology like the spear aided in killing big game. It is believed that hunters and gatherers eventually crossed the Bering Strait to North America and eventually drifted south and populated both North and South America. Those who remained in the Great Plains of the U.S. remained hunters and gatherers at least until the U.S. Cavalry came onto the scene.

When our country was in its infancy, there were estimated to be 30 to 60 million buffalo roaming across the continent with the majority living on the Great Plains. They knew no boundaries and as our country grew, they became an American icon. They appeared on stamps, the nickel, and in 1910 on the $10 bill. They became a powerful symbol of American history. They provided nearly all the necessities for the Native Americans, including

2 Changes In the Good Life

food, hides for shelter, bones for weapons and tools, and chips (dung) for fire. Throughout history during warfare, starving the enemy was a primary tactic. While attending Loras Academy, I took advantage of their ROTC program where I learned about these war tactics. This was accomplished by encouraging easterners to come west and slaughter buffalo for sport. Buffalo hides became in big demand by eastern buyers for leather, making clothing, blankets, and rugs to name a few. Just as the Romans spread salt on their enemies' crops, Generals Sheridan and Sherman used the same tactic by killing the buffalo to starve the Indians. This tactic had great success as by 1880, an estimated 200 buffalo were left, thus eliminating not only the Indian's food but also their way of life. At the dawn of the 20th Century, there were less than two dozen buffalo left in Yellowstone Park. (I read of a more accurate account of 21 head.) Besides these, some ranchers were raising skeleton numbers. Today's population is about one million.

Much has been written about the battle of Custer's Last Stand on June 25, 1876. Anyone who knows the story knows that the Northern Cheyenne, Arapaho, and Lakota won the battle but ultimately lost the war. The fact that these Indians were exclusively hunters and gatherers made this battle interesting and ultimately caused the death of Custer and the 7th Cavalry Regiment. Before the battle, Custer ordered Crow Indian scouts to case the Indian village along the Little Bighorn to see how many Indians were camped there. Upon returning, the scouts reported to Custer that there were too many and advised him not to attack. No one seems to know the exact number but historians estimate up to 5,000. Custer did not believe or take serious his scouts' report as to the numbers and Custer's mindset was that these Indians were hunters and gatherers and that a great many could not survive without being on the move to find food. Within two days after the battle, the Indians left their camp headed west where they could find food. Finding food became difficult as the buffalo were nearly annihilated. Before winter set in, most Indians had no choice but to surrender and live on reservations. For the most part this was the end of exclusive hunting and gathering in the lower 48 states. On June 25, 1876, the day of Custer's Last Stand, my grampo, John Kieffer, was four years and thirty days old.

The change from hunting and gathering to farming was a slow process as hunters and gatherers were spread out and constantly moving, going to where the food was present. Domesticating animals and plants was also a very slow process. Eventually after a few thousand years, the world became an agricultural community for the most part. However, there are an estimated five million people in various parts of the world such as Africa, South America, and the Arctic regions who live off the land or the sea. There have been and still are some people who are part hunter and gatherer and part farmer. I knew of a former distant neighbor and his wife, Willy and Kate, (now deceased) who would fit that description. They owned a 110-acre farm that was a long, narrow and hilly tract of land that was about one-half tillable. In today's world of agriculture, it would most likely not fit the description of a farm, but it was and still is a happy hunting ground. It has a creek going through it, woods, and all other features that wild creatures love and call their home. For man, it provided nut trees, wild apple trees, all kinds of wild berry bushes and fish in its creek, dandelions, and a good supply of watercress. Willy and Kate, lovers of nature and animals, spent most of their adult life on this farm enjoying and eating the bounty of food they were able to gather from it. Squirrels were one of Willy's favorite entrees as well as raccoon. Raccoon were a double deal for him as he could skin them, sell the hides and feast on the carcass. One fall Saturday, he invited several of his friends, including myself, to his house for a raccoon supper. Kate brought the raccoon from the oven and set it on the middle of the table with all four legs sticking straight up, ready to be carved. While Kate did the carving, Willy poured his guests a large glass of liquid spirits that he brewed earlier that spring. Eating raccoon was a new experience for me. When it was passed around, I wasn't sure if I would like it. I took a small helping and after trying a small fork full, I began to think my stomach my rebel. A slug of liquid spirit helped but I didn't know how I was going to consume the rest of what was on my plate. Will and Kate had a dog that had freedom to come in and out of the house at will. He had learned how to open the screen door. Willy and Kate had given him a unique name, Chicken Shit. Just as I was getting nervous again about my rebellious stomach acting up, Chicken Shit came in the house and for whatever reason came right to where I was sitting. He was the answer to my prayer. I was lucky that I was sitting at the end of the table so that while the others were busy eating, I was able to quickly and unnoticeably give Chicken Shit a good helping of my coon. I never ate raccoon meat again.

In the fall of the year it is quite common to see ducks, geese, and other fowl congregated together as though they are having a staff meeting about continuing their trip south. At this time of the year it is possible for some folks to practically live off the land as hunters and gatherers. This is especially true for those who have access to rural areas where wild game roam and nuts, apples, and berries are abundant. There is an abundance of all those foods on our farm. To some extent my family and I enjoy living off the land, but not to the extent of eating raccoon.

By 1802 Napoleon Bonaparte, a French military leader and emperor had defeated many European countries. Britain was his only major active enemy. To launch a full-scale battle against Britain, Napoleon needed money and a new and reliable way of preserving food. Negotiations to sell Louisiana to the United States for $15,000,000 were underway. The United States, at that time, was a young nation and like most farmers, wanted to expand their holdings. Britain wanted to sell as they needed money and finally a deal was struck in early 1803.

4 Changes In the Good Life

Can you imagine how challenging it must have been to keep an army of, say 100,000 troops fed in ancient times? Keep in mind there was no refrigeration and canning to preserve foods didn't come into practice until the early 1800s. Now these 100,000 troops could eat and would need to eat a massive amount of food to sustain the strength they would need to man their crude weapons and fight while being decked out in their heavy armor.

It was a common practice for troops to forage off the land and raid nearby villages, but this food supply would last only so long. A wagon train was employed to supply food when weather permitted, but note that rain made roads muddy and impassable. Troops soon began to go hungry and I recall from history classes troops would go mad from hunger before dying. Water was also a problem. The quality of water after having been carried in goat skins, and transported by camels for several days, seems to me to be quite gross. Another option for food supply was to bring livestock and poultry whenever possible. This seemed like the best way of feeding troops if the battles took place where vegetation flourished or if enough livestock and poultry could be obtained.

Moving forward to Napoleon, it was due to his push that preserving food by canning got underway. In the late 1700s, Napoleon offered a large amount of money to anyone who could invent a reliable way of preserving food. After about 10 years of research, a Nicholas Appert discovered that cooking food inside an airtight jar did not spoil as long as the seal stayed intact. Appert won the reward and preserving food by boiling in airtight containers became popular by the 1800s. There was still more experimenting that needed to be done as some foods responded well while others still spoiled. Between 1850 and 1860, Louis Pasteur discovered that bacteria caused the food spoilage and that heating the food killed the bacteria and kept it free of bacteria as long as the jar or can did not allow the bacteria to reenter. These discoveries led to easier and better ways of preserving food with the method of "canning." The name came to be because the first containers used were made from metal and called cannisters. Thus, the term, "canning." In 1858 John Mason invented the Mason jar which became the more popular container used for home canners.

Ann, my wife, kept up this canning tradition when our family was young, however she has given up much of it in recent years. I recall how she would put seven one-quart jars in a canner, cover the jars with water and boil them for 25 to 45 minutes depending on the food in the jars, then remove the jars from the canner to cool. As they cooled, the lids would 'pop' indicating they were sealed air tight. She would can tomatoes, tomato juice, pickled green beans, dill, sweet, and bread and butter cucumber pickles, apple sauce, pears, grape juice, etc. Corn, berries, apples and such would be frozen.

With the invention of the refrigerator, the refrigerator car, air-tight containers, and dried fruit, canning became less important, however mason jars are still quite popular items in the department stores.

In the fall of the year, a mere two generations ago, our grandparents would have a root cellar full of orchard and garden products to last through the winter. This produce consisted of potatoes, carrots, squash (especially Hubbard squash), rutabagas, turnips, etc. Different varieties of apples and pears were also found in the root cellar. Besides all of the above, they would have an abundance of mason jars full of just about everything a garden could produce.

In grampo and gram's day, before modern technology, hard physical labor was for most people, the way to survive. Five meals a day was the norm; people would eat to live not live to eat. Life is easier and better in many ways, but I think we have lost some good things along the way. We can't stop progress nor should we, but it is good to look back now and then to see if some of the good might be preserved along with the progress.

Gardening and preserving the produce today, for the majority, is becoming a lost art. Many folks just don't have the time, know-how, or equipment, and if they had all of this, wouldn't have the space to store it or the space to grow it. On top of this, it is far easier to go to the local grocery and buy it. As for me, gardening is still the in thing as I can't find fresh garden produce in the grocery stores that come close to the taste and quality that come out of my garden. Many of the jobs kids did on the farm have disappeared, but helping with the gardening and preserving the produce can still be a family project that produces bonding of families and pride in their accomplishments.

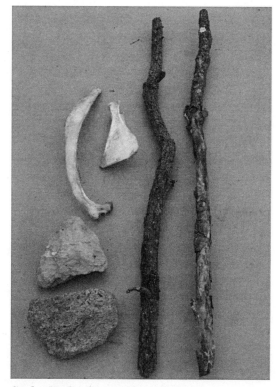

Sticks, Rocks, Bones, Man's first tools

Chapter 2

Evolution of Growing Grain

Part 1: Wheat

Bread has been a main staple food for thousands of years. Until primitive man started domesticating wheat, about 10,000 years ago, harvesting wild wheat was a slow and tedious process. It is believed that wheat was growing from the beginning of civilization. Triticeae is a wild grass from which wheat, barley, and rye originate. The process of taming wheat may be looked upon as an agricultural revolution as it made great strides in increasing wheat production, the primary ingredient in bread. Bread then became a primary food staple.

Wheat harvest, all hand labor

8 Changes In the Good Life

Domestication of wheat played an important part in the transition from hunter and gatherer to farmer. It was also the beginning of primitive people congregating together, living in villages or communities.

After man domesticated wheat, he continued to increase production by selecting the large seeds for planting, keeping the fields free of weeds, and selecting the best areas in which to plant. The practice of increasing production has been going on ever since by farmers and scientists. Different types of wheat have been developed, mainly spring wheat and winter wheat. Besides these that are classified by their growing season, they are identified according to their hardness such as hard red winter, hard white winter, durum, hard white and soft white. The hardness of the wheat determines what it is used for. For example, durum is the hardest of all wheats primarily used for pasta products such as macaroni and spaghetti. In the United States, durum wheat is primarily grown in North Dakota and Eastern Montana.

"I'm as corny as Kansas in August" is a line in the song, "A Wonderful Guy" from Rogers and Hammerstein's musical "South Pacific." I believe Rogers wrote the lyrics and I question if he had traveled through Kansas or Iowa before doing so. Had he done so, he should have noticed that Iowa is much cornier than Kansas. There is some corn grown in Kansas but wheat is the dominant crop as the arid weather makes it more suitable for wheat than corn. While it is officially called the "Sunflower State," more wheat than sunflowers are grown. Besides its reputation for wheat, Kansas from 1854 to 1864 was referred to as Bleeding Kansas which was triggered by the Kansas-Nebraska Act of 1854. This allowed, among other things, the settlers to decide whether or not to allow slavery. Also, there was the largest maximum-security penitentiary in the U.S. from 1903 until 2005 in Leavenworth, Kansas, bringing more notoriety to the State.

When talking about Kansas, I find it worth noting the lawless-frontier town of Dodge City. Lawmen such as Wyatt Earp, Bat Masterson, and a few others became famous for cleaning up the city of some of its law-breaking citizens, resulting in fewer bodies going to Boot Hill. I also find it worth noting Matt Dillon, a fictional character featured in TV's "Gunsmoke," put Dodge City once again in the spotlight, along with Miss Kitty Russell and the Long Branch Saloon and all the other colorful characters of the show. In more recent times, the small town of Holcomb in western Kansas became known nationally and in Europe for a gruesome murder of a wealthy wheat farmer, his wife and two teenage children. Truman Capote told the story in his book, "In Cold Blood," first published in 1966 which later became a movie. I read the book but I never had the courage to see the movie. I wouldn't recommend parents to read this book as a bedtime story to their small children.

Wheat planted in the fall is known as winter wheat. Due to the dry climate, winter wheat flourishes as it receives sufficient moisture throughout the winter to grow and mature. Harvest time begins in May in the southern states of the Great Plains and ends in late fall in Canada. At the end of harvest in a good growing year the elevators are overflowing with wheat. Traveling through Kansas or any of the wheat belt states, one would notice that the tallest buildings in small towns are grain elevators, filled to capacity with wheat by the end of July. The wheat is shipped from these country elevators usually by rail to larger terminal elevators in cities like Kansas City, Wichita, or Minneapolis, waiting for millers or foreign buyers. By the early summer of the following year when the rains diminish, the wheat is headed out and waiting for the grain to dry when the combines will come for harvest. The early pioneers tried unsuccessfully to plant wheat in the spring. It simply didn't work as the hot, dry weather set in. I would guess those farmers related well with the parable in Luke 8:4–8 from *The New American Bible*, "... *A sower went out to sow his seed. And as he sowed, some seed fell on the path and was trampled, and the birds of the sky ate it up. Some seed fell on rocky ground, and when it grew, it withered for lack of moisture. Some seed fell among thorns, and the thorns grew with it and choked it. And some seed fell on good soil and when it grew, it produced fruit a hundredfold." And after saying this, he called out, "Whoever has ears to hear ought to hear."*

Recently, I was at an agricultural seminar where the keynote speaker was talking about the many changes in agriculture. He said, "One farmer in Kansas sitting in his air-conditioned combine can harvest more wheat in one hour than all of Pharaoh's slaves could harvest in a day." I'm not sure where he got his information or how accurate a statement he was making about Pharaoh's slaves. I had never heard that so I did some research but came up empty. I suspect the speaker was using a figure of speech to make a statement about the great changes in agriculture.

Up until 1850, when Cyrus McCormick invented the reaper and later the threshing machine, changes in wheat production and harvest were slow or almost at a standstill. Hunters and gatherers simply picked the germ off the stalk. When man started domesticating wheat, he also came up with a simple tool, a sickle, to cut it. Soon the scythe came into use followed by the cradle. Animals walking on the wheat stalks with grain was one way of threshing it. Another was by using a Flail. Man's ability to grow and domesticate wheat made a great contribution to the end of hunters and gatherers.

In the mid-nineteenth century, with the coming of the threshing machine and the steam engine gaining popularity, the process of separating grain from the stalk or threshing became much less dependent on muscle power and much more efficient. When horses were used before self-propelled combines, harvest was a hot, dirty, and dusty job— hard work and very long days. When a farmer threshed his wheat, it took all of his neighbors, their horses and wagons, and their wives to prepare the meals. Any kids old enough to do anything were also recruited as there were numerous jobs for them such as carrying water to the threshing crew, driving a team, helping the ladies cook and the list goes on. When my Uncle Nick Tranel moved to Big Springs, Nebraska, using horses and a threshing machine was the in-thing. I learned much about the wheat harvest from him and his son, Lavern. The most important thing I can remember is how they rejoiced when the combine came with air conditioning. Goodbye to harnessing the horses at 5:00 A.M. every morning.

Like most changes, or so-called progress, something is gained but somethings are lost. With the coming of the combine, a great labor-saving device, started the end of the farm community working together. Threshing grain with a threshing machine and other such farm community projects such as silo filling, although hard and dusty work, brought farmers together to form tight communities of people who had fun together as they worked together, trust and interdependence among them. The men would gather at the house before each meal for a cold glass of lemonade, or perhaps a cold beer. In our part of the country, Pabst Blue Ribbon that came in a case of 24 small bottles was the choice. One of the first duties of the women was to make sure there would be cold drinks for everyone. The ladies would join together to socialize as they prepared the noon and sometimes the evening meal with a lunch of sandwiches and coffee or lemonade in between the two meals. Their next job would be to slaughter and pluck the chickens if that was to be the main course of meat. The meals usually consisted of chicken or roast beef, large bowls of mashed potatoes, gravy, coleslaw, and as threshing took place in July, numerous other garden vegetables were served. Of course, homemade pies would be the dessert, usually fruit pies at threshing time and very often pumpkin pies would be in season for the silo filling. Another job would be to set up a place for the men to wash up outdoors when they came for meals. A basin, towels, a bar of soap and a tub of warm water would be set up on a table under a large, majestic oak tree that could be found in the yard of the two-story house which served most often as the country home.

While there wasn't a lot that young kids could do to help with the threshing operation, very often some boys would come to be with their friends and they would be sent to do "gofor" jobs that they were able to do. Girls would also often come to help especially with taking lunch out to the field or whatever they could help with.

In those days, farming was definitely a neighborly, family way of life more than a massive business as we know it today. This is probably much the same as most family businesses of old, such as the family-run grocery store, hardware store, feed store, etc. It is one of the downfalls of progress but since progress cannot be stopped nor would we want to stop progress, it is imperative that we look for other ways to keep families and neighbors united through our church, school, and neighborhood communities.

When self-propelled combines came on the scene, the practice of custom combining began. Many wheat farmers, especially the ones with smaller acreage, found it impractical to own one of these gigantic and expensive monsters. It would be much more cost effective to hire their wheat combined. Owning a combine would be a bad investment as it would be used only a few weeks out of a year. So, it was the coming of the combine that triggered the practice of custom harvest. This also provided a good source of employment for college students. During the summer months of combining, there is much need for combine operators and truck drivers.

Today, wheat is high on the list of being one of the most important food staples of the world; corn and barley are also high on the list.

To finance our five kids (Ted, Tim, Pat, Tricia, Dean) through college, Ann and I devised a plan that we thought would help them financially while at the same time teach them to be responsible for finances and give them the satisfaction of some independence and make it easier for us as well. Our plan was that we would pay one-third of the cost (after deducting any scholarships they could receive), they would work for one-third, and get a student loan for one-third. The ideal work force on our family farm was myself and two full-time summer helpers and two part-time helpers such as high school students, during the spring and summer. When Ted graduated from high school, we were getting excess labor, not that there wasn't always work to be done, but paying a full wage was not a practical way of managing our operation. Shortly, before graduating he began thinking of a job and decided he would really like to go out West to find a job on a ranch for the summer months. He got ahold of the *High Plains Journal* and found a 'Help Wanted' ad in it looking for summer help on a wheat, hay, and livestock ranch near Liberal, Kansas. Of course, he got all excited thinking this to be a great adventure. He answered the ad, received a phone call, was interviewed and a few days after graduation was on his way to work on the Bob Hood Ranch near Liberal, Kansas. He soon found himself cutting an irrigated circle of hay. The summer went well and he got acquainted with a neighboring rancher, Dan Brown. Before Ted left Kansas to return to school in the fall, Dan asked him if he had any brothers like himself that would be looking for a summer job the next year. Ted came home with enough money to pay for one-third of his college expenses for that year. The next summer, Ted was back working on the Bob Hood Ranch, Tim was working on the Dan Brown Ranch, Tricia was working for her mother and Pat and Dean were working for me. That summer, Ann, Dean, Tricia, and myself visited Ted and Tim on their ranches. Pat stayed home to look after our farm. We found Ted and Tim happy with their work situation. A couple years later when Pat graduated, he got a job working in Iowa cornfields for an ag company and eventually, Dean had a similar job. Tricia worked during her summers in our local grocery store and for a daycare center.

Several years later, Ann and I were passing through the area and stopped to visit Dan and Anne Brown and their two boys, Thurman and Forest. Dan raises wheat in Kansas. He also purchases feeder calves to put on wheat pastures during the winter months. I asked about grazing problems:

Q. Is sickness in the young calves an issue?

A. We keep a close watch on them for the first few weeks and those with any signs of sickness are immediately treated. Once they become accustomed to their new surroundings, they start to grow. One normally has to watch the calves for 45 days before they really start to perform.

Q. Is mud a problem in warm and rainy weather; will they kill the wheat by tramping it in the ground?

A. Yes, if it gets too muddy, we have to take them off the wheat.

Q. I noticed a lot of the wheat fields do not have perimeter fencing. Do you use electric fences when you turn cattle into the fields?

A. Yes, but they can sometimes be a pain as the winds sweep across Kansas and Oklahoma at great velocities. The wind blows tumbleweeds against the fence that will knock the electric fence down. Sometimes tumbleweeds cause damage to permanent fences. It is also interesting to note that while these weeds are blown across the prairie, they are dropping a trail of thousands of seeds. God made sure that all plants propagated, even some we'd just as soon not have propagated.

Tumbleweed plants, are an invasive plant that are despised by ranchers and farmers. They are much like the water hemp, pigweed, and mare's tail, in that they produce a massive number of seeds allowing them to propagate rapidly, despite the farmers' and ranchers' attempts to control them, consequently giving Weed Scientists job security.

Q. What about rattlesnakes?

A. We have enough. Anne, Dan's wife, proceeded to tell a tale unlike any tale I had ever heard, about how their two cats ganged up on a rattlesnake. She clarified that the cats were outside or barn cats, not house cats. The two cats upon encountering a rattlesnake coiled up and ready to strike them devised their counter attack rather than running away. They moved closer cautiously, one to the front and the other to the rear of the snake. As the snake threatened to strike the cat in front, the cat in back struck with its paw the back of the snake's head. The snake would then turn his head to strike at the cat behind and the one in front would swat with his paw the back of the snake's head. This little back and forth game went on for a half hour or so as Anne watched from the house before the cats finally killed the snake. Dan proceeded to tell how rattlesnakes protect their heads, the most vulnerable part of their body. He went on to say that his protection from rattlesnakes is to carry a 6-foot soil probe to tap on their heads—two or three good taps on the head will do them in. He said that most folks try stabbing their head instead of hitting the middle section and then the snake will strike them or come up the shovel handle. Of course, an alternative would be to bring two cats along with you, providing the cats are snake smart.

Part 2: Corn

When primitive men started domesticating wheat in Western Asia, about 11,500 miles west of what is now Mexico City (the way the crow flies), the American Indians were starting to tame corn from a wild grass known as Teosinte. This plant is also known as maize. Today, it is one of the world's most important crops. It took thousands of years for this wild grass to develop into an ear of corn as we know it today.

On October 12, 1492, Christopher Columbus stepped on land for the first time after weeks of ocean voyage. A land that was full of strange people, cultures, clothing, and food. One such food he described as one of the general cereal or corn crops "which is a sort of corn which grows here, with a spike like a spindle." Exactly what he saw isn't clear, but certainly it isn't like what a Nebraska farmer sees today while driving his combine through a field in October nor was it like its origin, teosinte.

Bang board wagon

The corn people see today as they drive along country roads is derived from a long history of genetic manipulation. Most of this was done without really understanding what was being done to the plant. The result with each manipulation was a nicer looking plant that withstood the pests and weather, and produced more food.

The most primitive of today's corn ancestors is known as teosinte. Teosinte itself is certainly not a plant with many desirable qualities. The plant appears spindly. Tillers, or additional shoots terminating in a tassel, are prolific. When we see a tiller on a corn plant today, it derives from the main stalk at the base of the plant at the soil surface. Often it just looks like a runt plant. In teosinte, the tillers arise from the base of the plant as well as from the stalk a third or more up the plant.

Before I continue much more, it seems appropriate to talk about the structure of the corn plant. As with any plant, it has roots and a shoot. On the corn shoot, typically, a corn plant has about 18 leaves, although this can vary by five or more leaves. The leaves alternate the side of the stalk from which they appear. After the leaves appear, the tassel appears at the top of the plant. The tassel has the pollen. The ear eventually produces the kernels and is the part that most of us have interest in. The ear rises from where the leaves attach to the stalk. So, could a plant produce 18 ears? Not quite, but it could make quite a few. Not only do the ears come from where the leaves attach to the stalk but also the tillers. As the plant develops, tillers start to develop, although typically they don't grow much and we don't see them. Then when the plant has roughly seven leaves showing, give or take, tillers no longer start to develop from this location but instead ears start to develop. That's why teosinte can have tillers show from about a third up the plant and ears can appear above that.

Another unappealing aspect of teosinte are the kernels. Corn kernels have little leaves that surround them. With most corn today, these appear as dust while unloading grain from the combine to the wagon—sometimes called bees wings. With Teosinte these leaves surround the kernel and don't let go. Probably a good part of this is that the kernels are protected from insects and the weather to help in germination the following season. For trying to extract them for eating, it is not good. Once they get extracted, the next step is to crush them so they can be made into an edible food. Once again there is a problem. Teosinte kernels are very hard and difficult to crack. Once cracked they can be ground for making tortillas. With all of these problems, it is a wonder that people used this plant. And even more curious is why people selected this plant for improving. They did. What they did initially, no one really understands. And where they made these selections is even up for debate. The common thought is that the corn that the Pilgrims produced came from selections done by those in Central America. With the many documentations of the Mayan corn god and Inca drawings, this seems to make sense. However, some claim that there were corn drawings by Egyptians. Did corn derive from Egypt, then get transported to Central America many centuries ago? Did they rise separately? Historians still have some work to do.

We do know that corn appeared in a recipe book around 1570. Pope Pius V had a chef. Chef Bartolomeo Scappi wrote a book with the first known inclusion of corn in Europe. In the 1600's there are also records of corn in China. So, wherever it derived, shortly after Christopher Columbus sailed four voyages on the ocean blue it was everywhere. For those who visited a state fair to look at the biggest pumpkins or largest boar, you probably saw the tallest corn plant. Sometimes what contestants will do is use corn from Central America. My son Tim served in the Peace Corps in Guatemala. One of his many contributions was creating storage so that the corn they harvested would not be damaged by pests or deteriorate from molds. He also brought some seeds home which we planted. No, these never produced an ear, but they certainly grew tall. I probably should have entered them at the Jo Daviess County Fair in far northern Illinois. So why do I talk about my missed opportunity for a blue ribbon? Because if Columbus would have taken seed from his trips, they probably would not have produced much corn in many areas of Europe. If John Cabot, who landed on Newfoundland in 1497, would have taken seeds back, their success might have been better. The corn in North America in the early 1700's is what many of us would think of when we talk about Indian Corn. The multi-colored grain on a plant that may be green, purple, or somewhere in between was a staple for several of the groups of people in North America at that time. All the colors are indicative of all the various nutrients present.

Why did the kernels of corn that Tim brought from Guatemala grow so tall but not produce ears? First, we must understand florigen, which is a molecular signal in plants that initiates flowering. Plant leaves can measure the length of day. Based on the day length, plants can determine at what point they are in the growing season and, therefore, the right time to flower. Florigen is produced in leaves when the correct day length occurs, and then moves up the stem to the growing point to initiate flowering. However, day length changes across latitudes, so when a plant is moved to a latitude that it is not accustomed to, it is not able to determine the correct time to flower. The corn adapted to Guatemala did not initiate flowering when grown in Illinois, so it grew taller and produced more leaves but no ears.

Then in 1779, Lieutenant Richard Bagnal, while on a campaign, obtained yellow corn from the Iroquois. Yellow corn with various levels of hardness or flint to soft starch texture is the most common version today. Some variants of different colors still persist. From this time on, farmers continued to select the best plants. They would harvest, select the biggest ears, and use those for the following season. They planted in hills, or what might appear as a checker board (or chess board depending upon your preference) pattern from the sky.

This was done so that weeds could be managed. By taking a horse down the row, about 42 inches since that was a width of a horse, with a tillage implement behind, weeds could be removed. Then again taking the horse down through the rows but this time in the opposite direction, the rest of the weeds, or at least most of them, could be removed. With 3–5 plants per hill, they wanted plants that would not tiller, produce big ears, and withstand a little competition. The other thing they wanted were plants that would withstand diseases and insects. One way of protecting the grain from insects was developing husks that stayed wrapped around the ear. However, if the husks were wrapped too tightly, it would be difficult to hand harvest. By walking through the field with horse and wagon behind, having husks that would open easily with a tool such as a husking peg, and then allowing the ear to be easily snapped from the plant was a necessity. There was still wood to chop for the winter and numerous other jobs to be done before winter set in.

There was a slow and fairly steady gain in corn yields. Although we could just as well say that there is a slow and fairly steady erosion of the Rocky Mountains. One famous line of corn was called Bloody Butcher due to its deep red kernels. Robert Reid and his son James produced Reid yellow dent by selecting kernels from plants and ears that they liked, planted them and reselected kernels from the newly grown ears. George Krug and others followed. When selecting corn lines, the farmers wanted a few traits. Of course, they wanted yield. They needed plants that stood well with husks that limited insects getting to the grain since plants were planted in hills instead of rows. They also wanted ears that could easily be snapped off.

Then something changed. George Shull and Edward East discovered that when taking the seed from a plant that had pollinated itself, an inbred, growing that seed out, and crossing it with a different inbred, the resulting hybrid produced a bigger, better plant. The dawn of hybrid corn was born. With this the increase of yield went from minimal gain to just under one bushel per acre per year from about 1930 to 1960 to just under two bushels per acre since. More recently the integration of genes that allow the plant to resist insects and diseases as well as selecting genetics through molecular techniques has increased the rate of yield gain. It is not just these that have improved yields; it is also the improvement in fertilizers, herbicides, fungicides, and insecticides. Equipment that allows irrigation and fertigation, precise seed placement, uniform application of inputs, improved soil health with different tillage methods, and more has made the most of each kernel planted.

So, the question is what allowed for more yield? The kernels are about the same size. There are about the same number of kernels per ear. The main change was more plants per acre. This was allowed by the plants having greater resistance to stress from the environment and neighboring plants. Interestingly corn in China increased yield, but those increases have been due to more kernels per ear. Partly this is due to the continued use of hill planting in the more manual system in China.

With all of these changes it seems the life of the farmer would improve. Financially the price of corn did not keep up with inflation as it was just over $1 per bushel in the 1950s to currently around $3.50 to $4.00 per bushel. Instead of hearing the rustle of the corn and thuds of the ears against the bang board, farmers hear the purring of the combine engine and churning of the auger into the hopper and again into the wagon. Being able to sit while harvesting certainly would be considered an improvement. Having lights enabling work late into the night may not. The struggles of trying to figure out what genetics to plant, what new management practices suggested by salesman and university extension personnel, and is the soil dry enough for planting remain. While an expensive finger-pickup-32-row planter is a tough decision, trying to decide if one could actually pay for an expensive Oliver 402 four row corn planter wasn't any easier. Fortunately, another thing hasn't changed. After the struggles of planting, getting weeds controlled, and ensuring fertilizer is applied, the corn will start to shade the weeds. The shading is maybe a bit better now, but it is a time in mid-summer when a farmer can enjoy a slight reprieve from worrying about the corn. The site of a harvested field provides even more relief.

Part 3: A Brief History of Corn Planters

The first corn planter was a sharpened stick used by Mayan Indians of Guatemala and other Central American Countries when they started domesticating corn at about 2500 B.C. They would jab the stick into the ground to make a hole, drop in five or six or more seeds into the hole, cover it and move on to the next spot. They carried a handbag which rested at the waist with an extra-long strap around the shoulder to carry the seeds. The handbag might be hand woven by the farmer and/or wife or bought from a local rural weaver. The fiber and dye for the handbag sometimes is from the maguey plant growing in or near the corn field. Some still use that practice of planting corn on terraces in mountainous regions of Central America today.

My son, Tim, joined the Peace Corps in 1987 following his college graduation. One of his tasks was to work with the Mayan Indians and teach them more modern farming practices. He found it challenging as modern machinery could not be used in the mountainous regions that require all hand labor. Tim encouraged the Mayans to adapt to using the hand planter or the jab planter, but they were opposed to the "high-tech" tool. Planting corn was a festival and religious event that they

Man's first corn planter

18   Changes in the Good Life

didn't want to give up, which you can't help but respect them for, so they still use their sharpened sticks to plant corn on those mountain hillsides.

Stick planters are one of man's oldest implements. I would define them as a long, sharpened stick, long enough to suit the operator. They were easy to maintain as they could be sharpened with a bone, rubbed on a rock, or in more modern times, sharpened with a knife. The cost to own one was zero as they could be found under any tree. One drawback was that they required much labor. Labor was intense and the planting progress was slow.

Hand Planters

Using stick planters was the norm for planting corn seeds until the 1850s when hand corn planters made their debut. Numerous manufacturers produced thousands of these planters of many different styles. The one that I am most familiar with is the one in my antique collection. It consists of two boards three inches wide by three feet long with a metal (or tin) plate on the bottom of each board. A handle was attached on top of the boards. A seed-corn box measuring three inches by three inches by ten inches is attached to one of the boards. When the boards are spread apart, the metal plates could close forming a "V" that was poked into the ground. At the same time, seed corn would drop from a flat bar that slid out of the bottom of the seed box. This bar had an oval-shaped hole that could be adjusted with a

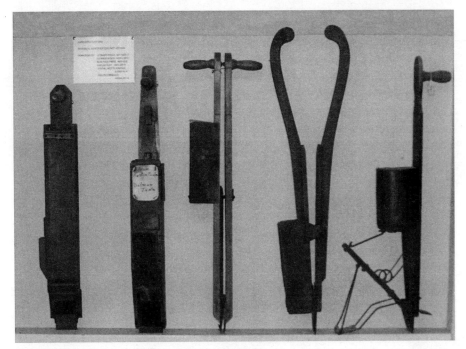

Hand planters that made their debut in the 1850s

clip and screws to determine the number of seeds to be planted. Next the farmer would close the boards, the bottom plates would open and seed fell into the ground. Closing the boards would also cause the bar to slide back into the corn box and reload, ready for planting the next hill. As the hand planter was moved to the next spot, the farmer would step on the previous hill to firm the soil. Of course, the farmer would step as heavy or light as needed based on his expertise and experience on the need of the seed to soil contact for his particular land. These are still used occasionally in small plot research when planting certain plots. An experienced farmer could move at a slow, methodical walk.

One-Row Planter

There is limited information about this planter but I have included a picture of it below. Apparently, a John Deere unit was first patented in 1882–1885 and a newer patent in 1901.

Row Marker

With the invention of the horse-drawn cultivator, it became important to plant corn in straight rows across the field in one direction and then at a right angle evenly spaced, creating a checker-board pattern. Corn was planted at each intersection with a hand planter. To obtain this pattern a row-marker sled was used. A simple, homemade device with four runners 42 inches apart that made tracks when pulled across the field by horses one way, then perpendicular to the first tracks across the field. The corn field could then be cultivated in either direction in the field.

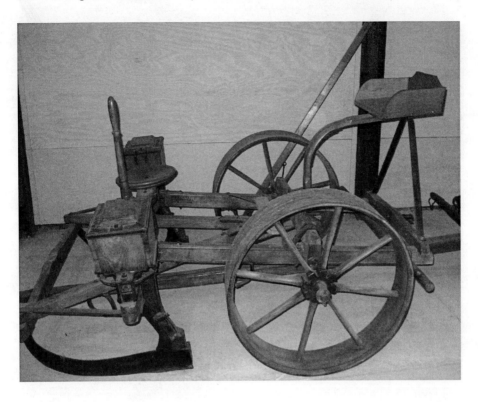

Hand Drop Planter

The hand drop planter developed in the 1850s was a major labor-saving implement even though it took two operators, one to drive the horses and one to operate the toggle stick. The row-marker sled was still used but it required only one pass perpendicular to the travel of the horses and planter as the planter had a guiding marker on each side that made a track that the horses straddled. As the planter crossed the line, the toggle operator would pull the toggle stick releasing seed from the box. Some farmers opted for a different way of knowing when to pull the toggle stick such as painting a stripe on the wheel or tying a rag around the wheel. When the stripe of paint or the rag touched the ground, the stick would be pulled.

Farmers took great pride in planting straight rows, both parallel and perpendicular. If those lines weren't straight, they would hear about it at the general store or after Sunday church services.

The Check Row Planter

The check row planter, patented in 1887, was used on our home farm as I remember well when I was a young boy. A detailed description appears in Chapter 10 so I won't repeat here. Note in the picture on the next page, the stakes used at each end to hold the wire, the planter plates hanging from the

seat and the roll of wire to the left of the left wheel. Before wire, a rope was used with a knot tied in it every 42 inches.

Modern Corn Planter

Today's planters feature row spacings varying from 30 inches, 20 inches, and some 15 inches. The most common row spacing is 30 inches. GPS eliminates the need for markers. Computer controlled systems sow seeds at a precise distance regardless of seed size, the feature most sought after by farmers. The planter has a bulk hopper, holding up to 50 bushels of seeds. From this hopper, seeds are blown to the furrow of each row. With the popularity of minimum or no-till practices, blades are mounted in front of each planter row so a furrow can be made.

Planting corn on Carroll Farms in Brazil

Chapter 3

Agricultural Industrial Revolution (1750–1850)

When we think of Agricultural Revolution, our thoughts might well go to the British Agricultural Revolution from about 1760–1850 when a series of innovations shifted, such as the textile production's change to a factory system. As a result of this change, cotton became the biggest non-food agricultural product. It changed the lives of people forever. Before the change, most manufacturing was done at home in rural areas with an entire family working together to produce products that were most in need. While that system was by no means as efficient as the factories of the future, it certainly kept families more united and less than 10% of the people of Europe were living in cities. Cotton production dramatically increased from essentially nothing in Mississippi in 1800 to over 535 million pounds by 1860. The vast majority of Midwest agriculture exports went down river through this nation's premier port city of New Orleans. By 1825 the Erie Canal was completed, and New Orleans quickly lost its monopoly on agriculture exports. While Great Britain still obtained most of their cotton for their textile industry through New Orleans, the flow of Midwest agriculture goods on the Erie Canal through

Steam engine and thresher

New York City changed the political and economic landscape by the mid-1800s. During the agricultural revolution period, scientific methods in crop and livestock practices were developed as well as the invention of more efficient farm machinery. One such machine that contributed greatly to man's food supply was the horse-drawn cultivator invented by a Jethro Tull, who also put it into use as a drill for planting seeds. Other machines that came into use were the steam engine, Ely Whitney's cotton gin in 1794, and Cyrus McCormick's reaper invented in 1831, even though the reaper didn't come into general use until about 1845. One could argue, or at least I will argue, that the horse-drawn reaper was one of the greatest labor-saving machines of all. It replaced the scythe which had been used for centuries for cutting grain, mowing hay, cutting weeds or brush, or just about any other plants that needed cutting. A person didn't need much training or brain power to run a scythe, but a great deal of brawn and ambitious perseverance was necessary. I know from personal experience growing up on the family farm. We had a horse-drawn mower to cut the hay fields and most weedy and grassy areas around the farm. Still there were many places that had to be cut with a heavy, cumbersome scythe such as around the buildings, the garden, orchard, and such. We would allow the sheep and goats to do a lot of "scything" wherever possible. My sisters would frown on having these creatures in the house yard but if we could find an easier way of cutting our workload, we quickly jumped on it. I would spend a lot of time sharpening the scythe on the grind stone but it always seemed to be dull. It took a mighty high-powered swing to make a dent in a grassy area. Sometimes, I would make two or three swings in the same place and the grass would still be standing there looking as if it were laughing at me. I suppose it could be more tolerable to use if building body muscles was your objective. It had an advantage over the present day weed eater as it caused no air or noise pollution making it much more environmentally friendly; it didn't use gas and it didn't get too hot, but it did build muscles. However, if a weed eater had been available, I would have thrown that scythe down immediately. I recall once seeing a railroad crew cutting weeds along the tracks with scythes on a hot day in July. I noted how they would make sure their scythes remained sharp by spending a long time under a shade tree sharpening them.

Now, if you think a scythe is a tiring tool to run, try using a grain cradle, that is, if you can find one to use. I have seen them in museums and antique shops but I'd guess no farmer kept them around longer than necessary. A grain cradle is a scythe with a cradle attached to catch the grain stock as you cut it making the tool even heavier and harder to handle. The grain stock would pile up on the cradle until the farmer laid it down on the ground in a neat pile where another person came along to tie the stocks into bundles. This is known as hand reaping which can still be seen in primitive parts of the world.

The first reaper was a rather crude machine that stood on two wheels and was pulled by two horses. The cutting part consisted of a steel plate or cutting bar. Steel points called guards were attached to this bar with rivets. The cutting was done with 3½ by 3½-inch triangle knives that were attached or riveted to a smaller bar about ¾ inch wide and ¹/₁₆ inch thick. This smaller bar with knives attached slid back and forth through the guards cutting the grain which became known as the sickle bar. The first reaper's cutting bar or head was six feet wide. It is interesting to note that the head on modern combines are up to 45 feet wide. The power to drive the sickle bar back and forth was obtained by the main wheel through a gear mechanism. As the grain was cut, it fell onto a platform and then had to be raked off into piles by a person walking behind and another person would follow to tie the grain in bundles. Improvements continued such as a self-raking device that raked the grain away from the platform which eliminated the need for a man to walk behind to rake the grain off. Soon after that improvement came the self-binder so now it was not necessary to walk behind to tie the grain into bundles. It did, however, require three horses to pull the machine with those improvements. A canvas belt was added that carried the grain over the main wheel to a box-like area where the grain would be gathered in bundles and wrapped tightly with string and a knotting mechanism would tie a secure knot. The result would be a neat bundle that was then automatically kicked out on the ground or onto a bundle carrier that would hold 7 bundles before the operator would trip a lever allowing the bundles to slide off the carrier to the ground. The seven bundles would then be picked up by people on foot to be placed together in shocks to dry. If the machine didn't have a bundle carrier, the bundles would be dropped individually so seven bundles would be carried (often this was a job for kids) to a spot where an older person would make the shocks. After the shocks were dry, usually taking about 2 weeks, the bundles would be loaded onto wagons and hauled to a threshing machine. To most simply describe a thresher is to say that it is a stationary machine that removes the seed of small grain from the stalks or straw. Small grains crops consist of wheat, barley, rye, and oats, etc., distinguished from larger kernel crops such as corn. Finally, the combine came into being and so named because it combined all the operations of the reaper, the bundler, the tier, and the thresher into one machine. The first tractor-pulled combine appeared in the 1920s. Twenty years later, self-propelled combines appeared and soon became the primary machine for harvesting small grain.

When horses were the primary source of power, before tractors, oats was a major crop which is not the case today. Oats was a high-energy feed for poultry and especially horses. A horse would consume about seven pounds per day, a pound or two more if they were doing heavy work such as plowing. The old-time farmers had it figured out that it took 1½ acres of oats to feed one horse for one year. Along with 1½ acres of oats, it took four acres of hay per year for each horse.

Next to the reaper in making the farmer's work load easier was the invention of the steam engine developed by Thomas Newcomer. In 1712 Newcomer invented the first practical and effective steam engine. His engine, as inefficient as it was, remained in use for about the next 50 years. James Watt began experiments on the Newcomer engine and made improvements. Watt took out his first patent on a steam engine in 1763. So finally, man had something more efficient than horsepower. The monstrous steam engine with its impressive horsepower was still not the answer to the average farmer's needs. They were extremely heavy, up to 30,000 lbs., making them prone to get stuck, especially in the mud. The biggest drawback was the danger of explosion. With that danger and the clumsiness of machine, they did not catch on to replace the horse on the farm. That would have to wait for the gasoline tractors.

In 1900 farmers had three sources of power—steam engines, draft animals, and their own strength. On American farms in 1910, there were approximately 24 million horses and mules. This was beginning to change with the commercialization of the combustion engine. During the first decade of the 20th century, combustion engines became hot item sellers. Farm chores started to become easier as these stationary engines took over many jobs, such as pumping water, churning butter, washing clothes, and so many more.

The first gasoline tractors had much in common with the steam engine. It weighed too much, was too big and clumsy, and cost too much. These tractors first went on sale in 1902. Technical improvements quickly got underway and continue to this day. First the size, weight, and price were reduced making it suitable and affordable for the average farmer. Henry Ford, along with his automobile business, introduced a small and inexpensive tractor model called the Fordson. Due to the shortage of horses during WWI (the demand for horses in Europe during WWI was much greater than the supply), tractors were in big demand. Ford's chief competitor was International Harvester Co. Those two companies got into a price war in the 1920s. Ford reduced its price for a Fordson from $625 to $395. International Harvester Co. had to reduce their price to stay competitive. Ford left the tractor business in 1928, returning in 1937. To compete with Ford, International Harvester (I.H.) Co. had to make great improvements in its tractors. The first was a power take-off. This was a major improvement as it allowed implements to be driven by the tractor engine rather than power from a wheel rolling along the ground. The power take-off became a standard feature on all tractors. In 1925 I.H. introduced the Farmall, a general-purpose tractor that was designed for not only plowing and soil preparation, but also for cultivating and harvesting. In 1927 John Deere invented the power lift. In 1927 Ford got back in the tractor business entering into a joint venture with Harry Ferguson. Ferguson introduced the three-point hitch, a device that was first used to keep a plow or a tillage implement level while traveling over uneven terrain. By 1932 rubber tires became available and by 1938 nearly all new tractors came equipped with rubber tires. By this time, draft horses were practically being stampeded off the farms.

Chapter 4

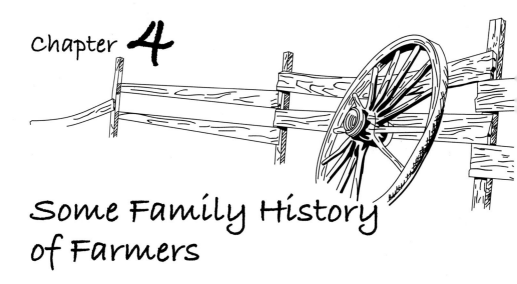

Some Family History of Farmers

As far back as I can trace my ancestors, they all made their living off the land. Farming must have been in their DNA, and still is, although in recent years many descendants journeyed on to what they saw as greener pastures. I am giving a brief genealogy as I am letting a more detailed family history up to those who are more interested in genealogy. Grandfathers are known by

KIEFFER STORE Menominee, Illinois 1913

1. Herman Kieffer(1st Lt.)
2. Hyacinth N. Kieffer(2nd Lt.)
3. Nicholas Kieffer(3rd Lt.)
4. Frank H. Kieffer(4th Lt.)
5. John H. Kieffer(Behind Boy)
6. Alphonsus V. Kieffer(Left Boy)
7. Joseph Kieffer(Tall Man)
8. Robert J. Kieffer(Young Boy)
9. Louis N. Kieffer(White Shirt)
10. Rose (Kuhl)Kieffer(White blouse)
11. Mary(Brummer)Kieffer(Older Lady)
12. Walter N. Kieffer(Baby)
13. Barbara(Kass)Kieffer(With Baby)
14. Mary L.(Kieffer)Little(5th Rt.)
15. Lucille B.(Kieffer)Tranel(4th Rt.)
16. August N. Kieffer(3rd Rt.)
17. Margaret(Portzen)Witterholt(Girl)
18. Blanche V.(Kieffer)Miller(1st Rt.)

Menominee General Store with Kieffer Clan

different names such as gramps, grampa, gramp, grandad, and grandfather. My grandfather was commonly known as grampo; therefore, I will be referring to him as such. I remember grampo as a thin man, 6'1" stature enjoying his rocking chair while smoking his curved pipe filled with dual purpose tobacco, Plow Boy, which could be used for smoking and/or chewing. He would be constantly puffing on his pipe. I never figured out if he inhaled or not as he was constantly puffing and blowing. When he smiled, one could note several front teeth were missing.

My great great-grandfather, Jacob, and his wife, Margarita, lived on a farm near Mulhback, Germany. Jacob was born January 1, 1819, the third of nine children. Jacob and Margarita had ten children, four of whom died under five weeks of age. Nicholas, my great grandfather, was the third oldest, born December 18, 1846 and died December 10, 1933. Jacob's brother Nicholas, who would be my great, great uncle, was the first to immigrate to America. My great great-uncle Nicholas settled in St. Donatus, Iowa, when my great grandfather was a young boy. (St. Donatus, IA is named after Saint Donatus the patron saint of good weather). When Great Grampo Nicholas was 9 years old, his uncle Nicholas (Jacob's brother) and living in St. Donatus, decided to make a visit back to his homeland in Germany. One can only imagine the conversations that took place back in Jacob's house. I'm sure there were great tales and many questions about America. Great Grampo Nicholas, a horse lover like most Kieffers, asked if there were any horses in America. His Uncle Nicholas' response was, "There are many horses and many wild horses just waiting for a young guy like you to tame them." The wild horse population in America at that time was estimated at about two million. Today, the estimated number of wild horses is 64,000. Apparently, my great-grampo did not forget those tales of America because when he was old enough, he also immigrated to America. Some family members thought my great grampo came to America to avoid the military but this doesn't seem to have much validity since this was a more peaceful time in the world. Neither my great grandfather nor my grandfather, John, nor my dad was called to serve in the military. Instead, I guess they threw baseballs rather than grenades and shot rabbits with 22 rifles instead of shooting enemy soldiers with Springfield rifles.

The National Archives in Washington D.C. has a record of the ship S.S. Cella that brought my great grampo to New York Harbor on July 8, 1869. It was an historical year in the United States as on May 10th, the president of the Central Pacific Railroad, Leland Stanford, used a silver hammer to drive a Golden Spike at Promontory, Utah. This ceremony signaled the completion of the Transcontinental Railroad by the Union Pacific and the Central Pacific Railroads. This was eight years and seven months after the Pony Express Riders made their last ride from St. Joseph, Missouri to Sacramento, California, with each rider carrying 20 pounds of mail in their mochilas. If all went well the approximate 2000-mile-trip could be made in ten days. The Pony Express started April 3, 1860, and ended on October 24, 1862, when it was replaced by the Transcontinental Telegraph.

When great grampo landed at New York Harbor much of the U.S. territory west of the Mississippi River was still in its infancy. It was a mere seven years since President Abraham Lincoln signed the Homestead Act into law. Over 1.5 million homesteaders took advantage of a practically free gift of 160 acres. Towns were being established and the railroads were encouraging folks to move west for greater opportunities. Large cattle ranches were emerging along with sheepherders and their flocks. On the horizon, there was a sea of grass, stirrup high, as far as the eye could see. Thousands of folks who had a passion for the land became pioneers in the Wild West. Others to go west were gunslingers, outlaws, and cowboys; a mixture that created a culture of crime before the arrival of the Law. Some who added to the Western Folklore included Wyatt Earp, Wild Bill Hickock, Buffalo Bill Cody, Billy the Kid, and Frank and Jesse James, who had a bad habit of robbing banks and trains. Buffalo Bill Cody was one of the most remembered as he was a soldier, buffalo hunter, guide, and later a showman. His shows were known as "Buffalo Bill's Wild West Shows" featuring Sitting Bull and Annie Oakley, who became known as "Little Sure Shot." These shows became big hits in the East as well as in Europe.

As the Wild West was growing wild, folks in the East and in Europe were yearning for wild tales about the West. Ned Buntline, a well-known author at the time wrote many western stories. Many stories about the West were published in Eastern magazines that sold well, whether they were fiction or non-fiction. Writing about the characters, culture, and color of the West was a 'gold mine' for both professional and amateur writers to make a quick buck. While all this folklore was happening in the West, my great grampo was content to settle in Menominee, Illinois.

From New York Harbor he headed west settling in Menominee, Illinois. There are no records as to how he made his way west. Now that the passenger train service was gaining momentum, all the way to California, great grampo could have hopped a train but riding trains were costly and great grampo's wallet was thin. In Menominee, he got a job working as a farm hand on the farm that he later purchased in 1871. On April 18, 1871, Nicholas married Mary Brummer who was raised on a nearby farm. From then on, the farm flourished and in seven years they had the 160 acres paid for. They had a reputation for keeping God's garden well manicured. In their spare time, if there ever was spare time, they would keep well informed of local and national news by reading the daily newspaper which Mary would walk one and a half miles to get. Can one imagine Mary having any spare time while raising ten kids and helping Nicholas do farm work, churning butter, feeding the chickens, gathering eggs, and the list goes on? These tasks were all done without today's modern conveniences. There is an old saying about farm work. A man works from sun to sun and a woman's work is never done.

Nicholas and Mary's first son, John, was my grandfather, born on May 24, 1872. Nine other children followed. Paul, the youngest was born March 13, 1893 making a span of 21 years between the oldest and youngest. Nicholas, Mary, and their children would occasionally visit relatives living near Bellevue and St. Donatus, Iowa. It was there that Grampo John fell in love with Anna Kieffer, a distant relative whose father was also named Nicholas. Her mother's name was Barbara (Ries) Kieffer. It is interesting to note that every Kieffer family had a Nicholas, Anna, John, and Barbara which makes following genealogy quite confusing. It seems the name Nicholas is the most popular. Catholic parents in those days would give their children a saint's name. St. Nicholas is the patron of children which is likely why the name became the first choice for many parents. Grampo John and Gramma Anna were married Oct. 23, 1894 and settled on a 283 acre farm near Menominee, which joined the one where he was born. Grampo John never moved far from home until 1959 when he and gram moved to a retirement home. Anna died at the age of 89 on Jan. 3, 1962. John followed her to Heaven a year later on Jan. 14, 1963. Their first child, a breech baby, died at birth in 1895, known only as baby Kieffer. My mom, Lucille, born Aug 24, 1895, was next and was followed by eight more children.

My mom and dad were married Jan. 7, 1920 and purchased a 137 acre farm for $20,000. Grampo signed a note for the down payment which Dad paid back in three years. Following WWI, land prices skyrocketed and remained high until the crash of 1929. My dad was able to make headway in paying down the balance of the mortgage in the glorious farming years of the early to mid-1920s. Their first born was Ralph on Dec. 27, 1920, followed by 12 more. I was the youngest, born at 2:00 P.M. on Nov. 2, 1937. The cost of my birth was a walloping $25.00. I was born on a beautiful Indian Summer day at home which was the norm in those days, especially for farm families. After my birth, I was laid on the kitchen table on a pillow made of goose feathers where I was put on display for my siblings to see when they got home from school. One comment I heard about later was made by 13 year old Richard, "He is so small, I don't think he will ever get big." Well, I proved him wrong as I grew a good five inches taller than he.

The feather pillow I laid on was made of goose feathers. Geese and ducks were the primary providers of feathers for pillows and mattresses. Geese were the best option since they were a bigger bird and yielded the better crop. My mom gave Ann and me each a feather pillow for a wedding gift. I never laid my head on a softer pillow. The synthetics are no comparison when it comes to softness. It took about three pounds of feathers to make a pillow and 20 pounds to make a mattress. Feathers could be plucked from fowl dead or alive. When geese would molt (shed their feathers) in the spring time, it would be the ideal time to pluck their feathers. The feathers would come off with minimum effort and cause no pain to the goose. When geese and ducks were killed for butchering, plucking feathers was a slow and tedious process as they had to be plucked in small bunches.

Chapter 5

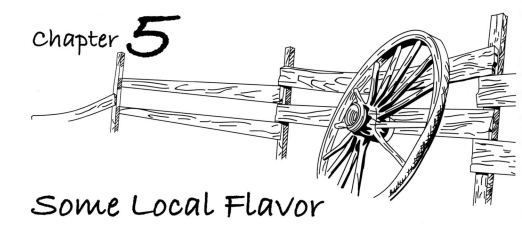

Some Local Flavor

By the early years of the 19th century, farmers from the eastern states as well as European immigrants were searching for land that would be more fertile than what they had. The migration to Ohio, Indiana, Illinois, and beyond began in full force. Today's rich prairie soil of Illinois was just what they were looking for. The rich soils of Illinois are some of the best in the world, comparable to those in Ukraine, which is known as the bread basket of Eastern Europe. A lot of blood was lost over the control of Ukraine as it was in big demand to feed troops especially during WWI and WWII. Initially the virgin soil of Illinois had little appeal to the migrating farmers as it lacked trees except along rivers and streams. The Illinois prairies also posed new problems with the prairie grass that grew up to seven feet tall. Even more troublesome is that grasses like big bluestem, little bluestem, and Indian grass had six-foot deep roots, compass plant and prairie dock had 14-foot deep roots, and stiff goldenrod as deep as 16 feet. These thick, tough roots made it difficult for a cast-iron

Breaking ground

plow (also spelled "plough") to penetrate the soil, and it was the best device available at that time. Even if the farmers could cut a furrow, they would have to stop every few feet to clean the soil from the plow. These problems existed until 1837 when the blacksmith, John Deere, made an improvement on the plow. He came up with a kind of steel used as a moldboard that turned the soil without the soil sticking to it. I think John Deere's invention was a great improvement to the plow and I wouldn't want to discredit him, but it has often been overstated when he is given credit for inventing the plow. In reality, his invention was the use of steel for the moldboard rather than cast iron. Turn the clock to 700 B.C. Isaiah is quoted in Chapter 2, Verse 4 of the New American Bible thus: *"He shall judge between the nations, and impose terms on many peoples. They shall beat their swords into plowshares..."* This passage would indicate that plows, probably a very crude type, were then used. When planting seeds started as I mentioned in Chapter 1, man used a stick, rock, bone, or anything else that could be used to loosen or turn the soil.

Since its invention, the moldboard plow has been the most popular. A moldboard is a part of a plow, a curved piece of metal designed to turn a strip of soil over or upside down. This strip of soil could generally be from 12 inches to 16 inches wide and six inches to ten inches deep. Looking across a freshly plowed field, it can look as soft as a mattress.

Another problem with the prairie land of Illinois was that it was full of wetlands and swamps. It wasn't until 1870, that tiling and drainage ditches got under way. This was a slow process as the digging had to be done by hand with a shovel. This had to be a back-breaking task.

Jo Daviess Co., located in the far northwestern part of Illinois, for the most part, did not have swampy land to contend with except for a few areas along rivers and creeks. Jo Daviess Co. is bordered by Wisconsin to the north and the Mississippi River to the west. The topography is quite varied, from steep cliffs along the river to rolling hills and some very rich virgin soil. Menominee Township, where my ancestors immigrated and grew up, was typical of the county's topography. The farms today on average consist of about half tillable acres with the rest being either pasture or timber land and many rivers,

Axe

Cross-cut saw

small streams, and ponds. Abundant pure, delicious spring water flows freely in numerous areas. My farmland is quite typical which I love as I enjoy the recreation and indescribable beauty it affords with each new season.

Trees with very large trunks were in abundance in Jo Daviess Co., just standing there waiting to be cut down with long, sharp, cross-cut saws and sharp, double-bitted axes. This helped to attract settlers to the area. Keeping saws sharp could sometimes be challenging, especially if they encountered a solid object like a horseshoe or an arrow head imbedded in the tree. I recall many years ago, a story told by my uncle Nick, about how he and my dad tried to cut down a gigantic tree. They just got started into the tree when they hit something solid. They started sawing again, after sharpening the saw, slightly above the first cut just missing the solid object. After the tree fell, out of curiosity, they used an axe to chop out the object, discovering it was an arrowhead. They concluded that many years ago a Menominee Indian missed his target, shot low, hitting the then-small tree, penetrating it deep enough that the tree grew around it and concealed it.

One evening, while eating supper, my older brothers were telling about hitting a horseshoe while bucking up limbs for firewood. The 28-inch diameter circle saw they were using was mounted on a platform on the front of a John Deere B tractor. The saw was belt driven from the tractor belt pulley to the saw shaft pulley. The saw had a wooden guard on its sides and top that went back and forth on tracks. This method of sawing would require five or six men (not boys as this was considered a man's job). Three or four men were needed to lift the long limb onto the platform, one to push the limb into the saw and pull the guard back after the cut was made, and the other job was to grab the cut piece, throwing it on a pile. This way of bucking wood was eventually replaced by the chainsaw, although not as fast but only requiring one person. Both methods were certainly a big improvement over the bucksaw. As I mentioned before, sawing wood took five or six men which was a neighboring project. The invention of the chainsaw eliminated another neighbor gathering or social event.

Hitting the horseshoe slowed down the operation as the saw had to be sharpened. My mom shuttered as she heard the story, visualizing a sawtooth breaking off and hitting someone. Fritz, the former owner of the farm, by now deceased, took a lot of flack for hanging the horseshoe on the limb where it eventually embedded.

34   Changes in the Good Life

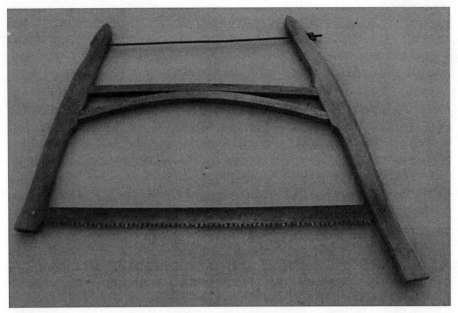

Buck saw

    The axe is one of man's oldest tools and weapons. The first was a stick with a bone or a wedged type rock attached to the stick. Eventually copper and then metal replaced rocks and bones. Crosscut saws came into use in the 15th century. The axes and crosscut saws were among the most essential tools for survival in the early days of our country.

    The settlers of Menominee Township and northwestern Illinois had a different problem on the tillable acres than those of the prairie region—clearing trees. This was a slow process taking decades to cut the trees, then grubbing the roots out with horses. Some of the hardwood trees consisted of Ash, Elm, Hickory, Locust, Oak, and Walnut. Oak was the most desirable for building livestock yards and used for fence posts due to its durability. The stumps of large oaks, too big to grub, would take decades or even centuries to rot. I have some oak stumps on my farm that I would estimate to be well over 100 years old. To grub a tree stump, a ditch was dug around it and most of the roots were chopped off. A chain would then be wrapped around it and hitched to a team of horses. The horses would attempt to pull it out, sometimes successfully and sometimes not. Anyway, this gave the horses some additional employment when they were not pulling a plow or binder or some other fatiguing job. Some very old, gigantic trees too big for the pioneer's saws were girdled and left to die a slow death. Timber was not only an asset to the settlers for building but it was also in big demand in Europe. Ever since and through the colonization of the United States, timber was and actually still is an incredible resource. It

Some Local Flavor   35

would seem that the Europeans have never valued conservation of their tree supply as they have been stripping their resource since Roman Empire days for use in housing, fire, and extensively in shipbuilding. It has been suggested that one contributor to the demise of the Roman Empire was the loss of wood resources. In the 19th century, Oak trees were in great demand in European wine making countries for use as barrel staves in their wine casks. It wasn't until the 19th century that Europe began making regulations to help replenish their timber supplies.

This timberland also was a great habitat for wild animals and birds, especially woodpeckers. As a side note, I found in my research an interesting bit of trivia about woodpeckers. They have a unique tongue as it is three times as long as its bill. It has two primary purposes—one is to dig out insects and the other is to protect the tiny brain of the woodpecker while he is pecking. I thought this is some valuable information that everyone should know, right?

At the same time that pioneers were clearing trees, they were planting fruit trees, following in the footsteps of Johnny Appleseed (John Chapman), especially apple trees. A small portion of apples were used for eating while the majority were for making hard cider, a popular beverage at the time. A decanter of hard cider was commonly sitting on the dining room table at meal time. Pioneers were used to making their own hooch until prohibition kicked in in 1920 at which time FBI Agents would go to orchards with crosscut saws, cutting down the apple trees without asking any questions.

Settlement began in the early 1830s in Menominee Township and Dunleith Township, part of Menominee, until 1865. Prior to European settlers, black bears, gray wolves, and mountain lions were a part of the landscape in the Midwest. As people populated the area, the population of those wild animals decreased to almost none. When settlement got underway in Menominee Township, there was, and still is, an abundance of wildlife consisting of deer, mink, coyotes, beaver, raccoons, muskrats, possums, foxes, otters, squirrels, wild ducks and geese, prairie chickens, pheasants, tons of rabbits, bald eagles, and other birds of numerous species. One animal I can't omit is the most damaging little varmint, the groundhog. Groundhogs like to dig holes all over my farm including around my farm buildings. Had Noah only thrown that pair overboard. But without Phil, the groundhog, who made his debut in 1887, how would people of that time and even today predict the coming of spring? I, personally, have little faith in his accuracy as I believe his predictions are correct less than half the time.

Menominee Township was settled by German Catholics and named for the Menominee Native Indians. The township has a village that was incorporated in 1940 with a population of 138, now 258. In its earlier days, the Village of Menominee had a post office, creamery, cheese factory, blacksmith shop, baseball field, a Catholic Church and rectory, a Catholic school and a convent for the Catholic sisters who taught in the school. It also had a grocery and general store built in 1800. The church, cemetery, and baseball field are still being used as well as the building that housed the grocery and general store, although as a different business. When my great grandfather, Nicholas, retired from farming, he purchased the store and in 1919 sold it to his son, Louis, who ran it for 30 years as a grocery and general store. In 1948, Richard Runde and his wife, Eloyse, purchased it and called it Runde's Tap. In 1979, a fire started, causing considerable damage and it sat idle until 1983 when the Runde's daughter, Karen and her husband Jerry Meyer took over, repairing it and naming it JM's Tap. Today, it is noted for its delicious fish fries, pizza, sandwiches, and a friendly place to gather in the community.

The village of Menominee has a unique feature. It has only one sidewalk that leads from the tavern to the church or does it lead from the church to the tavern? The Nativity of the Blessed Virgin Mary Catholic Church in Menominee was built in the 1870s and is made up of just under 200 families to occupy the wooden pews that make nostalgic creaks, reminding everyone of their sturdy longevity. Each person has their particular pew they usually sit on and when they are missing, there will surely be conversation as to why and bring someone to their home to check on them to see if all is okay. Church goers in this church, like many Catholic and Protestant churches, fill up the last pews first, often leaving several of the front pews empty waiting for worshipers. There are exceptions to this, like when one of the organists, Jerry Stangl, plays certain songs such as "Here Comes The Bride" or "Amazing Grace"—weddings and funerals.

In the horse and buggy days of my ancestors, the majority of people attended Sunday church services on a regular basis and in rural areas such as Menominee, this required great sacrifices, especially in the cold winter months. The farm families would have had to awake early to attend to their livestock, then hitch up the team of horses and travel for several miles on all kinds of roads in all kinds of weather. Today, we go to our heated garages, get into our warm vehicles, and travel mostly paved and cleared roads. Some vehicles have a remote starter, so by pressing a button before leaving church, we can start our vehicle and have it cozy warm or comfy cool. The walk from the parking lot to the church is probably our greatest discomfort. As I mentioned earlier, in the horse and buggy days, nearly everyone made the sacrifice to attend church services every Sunday. Today in spite of our conveniences, only about 30% of our nation's population bother to make Sunday services their priority.

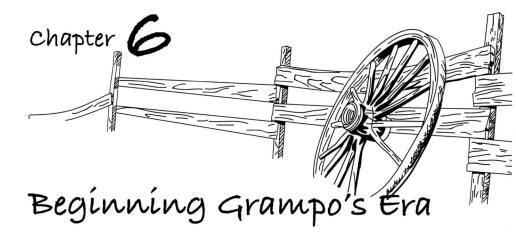

Chapter 6

Beginning Grampo's Era

When grampo was born in 1872, attending school wasn't mandatory. Not many graduated in those days. Students either dropped out to work on the farm or in the kitchen or because they missed so much school, they got too old to attend any longer. It wasn't until 1883 that the compulsory education law in Illinois went into effect. The law required children ages 7–16 to attend school. At that time, this law was difficult to enforce and many children continued to work, especially farm boys. The Child Labor Law in Illinois went into effect in 1893. This law primarily was to forbid children to work in factories. Many children working on family farms continued to skip school. In 1916, a law forbidding children to work in other occupations was passed and the compulsory school

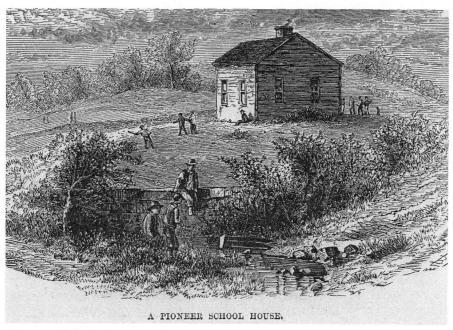

A Pioneer School House

38 Changes In the Good Life

law began to require more compliance. From conversations that I remember hearing from old timers, farm kids still didn't take going to school seriously and many parents didn't encourage education as they were so needed on the farm. I have also heard that truant officers simply looked the other way, especially in rural areas.

Wisconsin passed a compulsory education law in 1879. Iowa didn't pass one until 1902. My grampo was born 11 years before the compulsory education law and attended school only a few months during the winter, after the harvest was complete and before the spring planting started again. Kids were expected to stay home and work full time on the farm. My grampo, being the oldest child, received less formal education than any of his siblings, however he became self-educated and kept up on world events throughout his entire life.

To satisfy their social and entertainment needs, farmers and their city cousins alike would do a lot of visiting with friends and relatives, playing cards and other games such as board games, checkers, chess, etc. My great grampo Nicholas would visit his relatives in Bellevue, IA, whenever possible, taking his family along. This is how my grampo met and fell in love with his future bride, Anna Kieffer. They were related but distant enough to be acceptable. On Oct. 23, 1894, the marriage took place at Bellevue, IA. At the time of their marriage, grampo had a farm operation under way on a 283-acre farm that joined his dad's farm.

After the marriage, grampo brought his bride home, a distance of 32 miles, not in a fancy surrey with a fringe on top, but in a rickety old steel wheel wagon. This was probably the only wagon he owned which would have been used to haul his hogs, sheep, and other livestock to the East Dubuque stockyards or cream and other produce to the Menominee, IL or Dubuque, IA creamery. Selling cream was a once a week job.

The day before the wedding, grampo took the livestock racks off the wagon, hitched it up to a team of sorrel mares, his two best horses, and drove it to the Menominee River (which had more resemblance to a creek) that ran through his farm. There he parked in the river and gave it and the team a good scrubbing. Grampo knew the ride home would be bumpy but he at least wanted it to be clean.

To assist grampo in his early years of farming, he had a hired man, Paul Greenwalt, a bachelor. Paul was a good worker but he had a bad habit of taking a day off now and then whenever he pleased. This arrangement lasted until grampo and gram's kids got old enough to replace Paul. In 1895 their first baby died at birth as the result of a breech birth and the doctor did not arrive in time. Their second baby, my mom, arrived August 24, 1896. Eight more children followed providing grampo and gram with sufficient farm help and finally allowing Paul to take as many days off whenever he wanted to.

Grampo's farm was comparable to most farms in the area in his day with a couple of exceptions. It was larger than average with more pastureland and timberland than average. Another advantageous feature to it was its scenic beauty. Looking north, one could see the Menominee River flowing through the bottomland well suited to raise a bumper crop of corn providing it didn't flood, which happened about one year out of ten. To the northwest of this valley lies a large bluff that provided excellent habitat for bald eagles. They could flourish here with the river providing them with an abundant supply of fish, an important food source for bald eagles. Watching them make a downward flight of up to 100 mph to the river to snatch a fish would be a pleasurable sight for sure.

To the west, grampo could see the Herman Vogel Farm. Not only could he see his farm but he had an excellent view of the smoke that came out of Herman's hooch house when he would be brewing. The house itself was well hidden sitting about half way down a steep hillside. Herman Vogel's wife's brother and sister-in-law were killed after being hit by a train, leaving six orphaned children. Herman and his wife had two children of their own and after the accident the six orphans moved in with them. Herman Vogel's farm, due to its terrain, was better suited for hiding a hooch operation than it was for a farming operation. Herman put those hills to their best use, built a hooch house and a profitable operation of making whiskey to provide for his large family.

Herman would store the whiskey in milk cans so as to disguise the whiskey and whenever he had a sufficient amount to sell, the "cans of milk" would be transported to Dubuque, IA on a horse drawn wagon. It was widely known that he never got caught selling or distributing illegal whiskey. Legend has it that either the Federal Agents really didn't know of his operation or they knew what was going on but simply turned a blind eye realizing he needed the money and was doing a good deed by raising the six orphan children.

Beyond Herman Vogel's farm farther to the west, lies the farm where Eleazer Frentress plowed the first furrow ever plowed in Dunleith Township. Eleazer and Diadamia owned the first recorded tract of land in the area. Eleazer Frentress, 1800–1853, and wife Diadamia were parents to 13 children of which only six survived. They settled on 1,000 acres in the East Dubuque, IL area. They also had 600 plus acres near Beetown, WI, where Eleazer did some lead mining for a short time. In Iowa, they also established an 1,100 acre farm. He served in the Blackhawk War and died at an early age of 53 years. It was in the spring of 1832 when the Blackhawk War was all but suspended, that Eleazer commenced breaking the prairie sod with a plan to plant crops.

Through the ages, containing livestock was challenging. The invention of barbed wire was the greatest asset in containing livestock. As we now know, barbed wire was first patented on 11/14/1874 by Joseph Glidden of DeKalb, IL. Glidden's barb wire was not the first design, but it improved on others as it used two strands of wire twisted together to hold the barbed spur wires in place. Before barbed wire, various materials were used such as timber, stone, brush, or whatever native raw material was available. Smooth wire was tried but proved to be mostly ineffective. Building a fence with native raw materials was very labor intensive. On my farm, I have the remains of an old rock fence built between my land and a neighbor's. Since it was built it became useless as it was built in the wrong place. Just looking at this fence gives me a backache realizing how much work there was to building a rock fence with hand labor. The simple invention of barbed wire made a huge impact on life and culture on the farm and especially in the Great Plains. Like many new inventions there were downsides to barbed wire. The owners of large ranch operations were unhappy as settlers enclosed their 160-acre homesteads. This caused the beginning of the end of free range. Homesteaders at last had a simple and effective way to enclose their livestock. The relations between the farmers and cattlemen became tense. The cattleman would cut the farmers fences leading to range wars. Finally, the days of the open ranges were over. I could write much about range wars but that can be a story and a half for another day or year. The musical "Oklahoma" has a song about the tension between "The Farmer and The Cowman." A couple lines—"The Farmer and the Cowman should be friends" and "Territory folks should stick together."

Henry Frentress, (1842–1899) son of Eleazer and Diadamia, continued in the management of the Frentress Estate and also patented the split diamond barbed wire on 1/14/1875. This wire was manufactured in Dunleith, IL. This wire must have been made of high-quality steel as some of it can still be seen on a small portion on one of my fences.

To the northwest, grampo could keep a close eye on his daughter, Lucille (my mom), and his son-in-law, Ed Tranel (my dad). They lived on a 137 acre farm joining his. Farther to the northwest beyond my mom and dad's farm, grampo could see Ignatz, better known as Iggy, cultivating corn during the entire month of June, usually starting at 5:30 A.M. I remember Iggy as kind of an old guy who held onto outdated ways like cultivating corn with a team of horses long after tractors replaced horses. Most of his machinery belonged in a museum. Often following behind him would be his three-legged dog. Three-legged dogs were not such an unusual sight those days, as when farmers would be cutting hay with a sickle mower, dogs would be busy chasing rabbits and other creatures being frightened out of the hayfield. The dogs would get too excited, not watch where they were going and occasionally get a foot too close to the sickle bar resulting in a quick, unintentional amputation. Iggy's dog had healed quite well and kept up to anything that moved on the farm in spite of his challenging condition. While cultivating corn, when I was a young boy, my job would be to drive the horses while one of my brothers would walk behind guiding the shovels of the one row cultivator. I had the easier job as I got to ride although I was too little and too unskilled to guide the shovels. Iggy managed to do both jobs by himself. He would tie the lines together in the back of him with the right line over his right shoulder and the left line under his left arm. When he wanted to turn to the right, he would twist his body to the right and a left turn required a left twist of the body. This was a common way of driving horses when the operator needed their hands free to guide an implement such as a walking plow or cultivator. In reality, little guidance or turning the horses was necessary except on the ends of the rows as the horses would become accustomed to walking through the center of the rows. It's interesting to note that straight corn rows were a matter of great pride for farmers. Way back when in grampo's farming days, farmers had certain tasks that they wanted to have completed by certain days, such as having the corn cultivated by the 4th of July, better known as having it "laid by." Other goals were to have the first cutting (crop) of hay harvested and stored in the hay loft. By Thanksgiving corn picking should have been completed and wood should be stacked up to the rafters of the woodshed before the snow flies.

Some farmers supplemented their wood supply with coal, which reminds me of an interesting story. The longest uphill railroad grade between Gillette, WY, and Chicago, IL, is in Scales Mound Township of Jo Daviess Co., IL. In this township there is a gentle hill known as Charles Mound, rising 1,235 feet above sea level, the highest point in Illinois. This rail line is used to keep Chicago supplied with coal from the coal mines of Gillette. In the old days, if the weather conditions were less than ideal or if the train cars were somewhat overloaded, the steam locomotive would spin out, causing a stop. The tracks ran through this one farm family's farm on this uphill grade so they had a unique and inexpensive way of acquiring coal to supplement their fuel supply, although it could be dangerous and was most definitely illegal. They got to know just about where the locomotive would spin out and to make sure it stayed stuck they would apply a liberal coat of grease to the tracks. While the train was just sitting there with steam and smoke puffing out, all able-bodied family members hopped up on the coal cars, helping themselves, throwing coal off as fast as they could.

In the pretty valley where my Dad's and Iggy's farms joined, early in the morning you could hear the sounds of all kinds of birds chirping, rustling of leaves made by the squirrels playing in the trees, roosters crowing on a distant farm, frogs croaking in a nearby stream, or a stray doggie bawling for his mommy, yearning for his first meal of the day. These and numerous other sounds of God's creatures created a symphony in the halls of nature. There might be one exception on a few mornings of the year when a discordant sound could be heard. Just a country mile away Iggy may be out cultivating corn and while doing so, be constantly hollering and cursing at Prenty, one of the two horses he used to pull the cultivator. The other horse's name was Peanuts and Prenty would be lagging behind not pulling his share of the load.

Breaking Prairie

Winter Farm View

Early Hayrake—This implement slid along the ground. When front section was full of hay, it was turned over frontwards to dump the hay. The rear section then became the front and the process continued.

Threshing Machine—First appeared in 1786.

Yesterday's Corn Crib and Today's Corn Storage.

Old Barns—Empty barns full of memories.

Pig in a tricycle cart going to market in Brazil.

Prized Stallion owned by my father-in-law, Joseph Berlage.

Bang Board Wagon & horses

With this rig, one man could pick 100 bushels of corn in a day and scoop it into a corn crib.

Husking Peg

Shovel

Corn Picker—2MH cornpickers became one of farmers most prized implements following WW II. Pictured is Ross Klein, local IH dealer and Urban Wubben driving in 1963.

Grain Cart 4,000 Bu., 58' long, 13' tall, 14' wide, 22' auger.

Modern Combine—This one-year old John Deere, model #S780 combine, owned by Judy and Tom Schuldt, and Jeff and Heidi Schuldt, comes with a comfy cab loaded with multiple screens providing much important data and features 24 rows @15 inch row spacings. This monster machine can combine 2500 bu. corn per hour and can chew through 12 acres per hr. at a speed of 5.5 mph. It sucks out 22 gallons of diesel fuel per hr. from its 375 gal. fuel tank. It is able to harvest more corn in less than 5 minutes than one man could pick by hand all day, throwing ears into a bang-board wagon. The cost, as Tom mentioned, is $450,000 for the unit plus $140,000 for the corn head – just a tad more than the cost of a team of horses and bang-board wagon including the hand corn-husking peg. I rode with Tom in the deluxe cab for about an hour while he explained the operation's center as it recorded yield, moisture and much, much more.

Drone—Scouting a cornfield

Combining wheat on Dan Brown Farm in Liberal, KS.

Wheat—Golden wheat ready for the combine.

Planting corn on the Carroll farms in Brazil.

Chapter 7

Farm Buildings

Within a few years after gram & grampo married, they built a new house. This house was large with a furnace in the basement, radiators, and quite modern in its day, it had a white picket fence around it. Grampo's farm buildings were typical of most farm buildings of the early 20th century in the Prairie State of Illinois and other Midwestern states. Diversified farming was the norm; thus, many buildings were needed. Barns, the most important farm buildings besides the house, are described in the dictionary as an agricultural building used to house livestock as well as fodder, grain, and equipment. As a result, the term barn is often qualified as the sheep barn, tobacco barn, potato barn, cow barn, horse barn, etc. They came in an infinite variety of shapes, colors (although most are red), and sizes that are the heart of the farm. When early settlers came from Europe, they brought ideas about how a barn should be built and appear, but they had to make one major change which was to use the raw material available and that was timber which was in abundance. Thus,

Farm buildings in Grampo's Day

most barns and other farm buildings are built out of lumber. Much has been written about barns, i.e. the building material used, types of barns such as the Dutch, Bank Barns, Round Barns, Prairie Barns, just to name a few.

Farm buildings on homesteads in grampo's day were smaller but more plentiful due to the fact that farmers were more diversified back then whereas today specialization dominates the landscape. It was not uncommon for farmers to have numerous livestock and poultry, both beef and dairy cows. Shorthorn cattle was the preferred breed of the pioneers originating from England, valued for their milk and meat. Other animals of value that needed shelter consisted of horses, sheep, hogs, chickens, turkeys, ducks, geese, guinea, llamas, donkeys, and mules. Of course, every farm in grampo's day had more cats than they needed plus one to two too many dogs. With the above-mentioned animals, the need for many buildings is evident.

Besides livestock shelter other buildings were needed, i.e. granaries to store oats, rye, barley, wheat, and a corn crib. A workshop was a very important building on all farms. On the farm I grew up on, it served several functions. It ranked second to the hay loft in providing a great playpen for me and my siblings. We got on-the-job training fixing things, and developed carpenter and mechanic skills. Of course, in those days, much of your machinery could be repaired at home. Today with all the computer systems, a formally trained mechanic is required. We developed skills in creativity, which in my estimation is something one never forgets. There was certainly a lot of bonding that took place in that shop as we spent cold and rainy days working on some project or other. Our workshop was always stocked with a good supply of at least four kegs of six, eight, 16, and 20 penny nails. I have great memories of driving many nails in building bird houses, miniature barns, toys, and even a Christmas crèche for my mom to use at Christmas. And then there was the real work of oiling and repairing harnesses, bridles and halters, and greasing wagon and buggy axels. Sickles, scythes, and knives were always in need of sharpening. I don't think the to-do list hanging in the shop ever runs empty then or now.

Workshops in grampo's day were equipped with nails, bolts, screws, scrap iron, and vintage tools, but no power tools. All tools required muscle power, such as augers, vertical drills, anvil and hammers, a leather punch and rivets for harness repair. A good supply of leather was crucial to have in stock to repair harnesses, bridles, and halters. Hand saws and crosscut saws, wedges and sledge hammers would occupy a place in the shop, along with a grindstone and forge. The primary use of a wedge was to aid in falling a tree. It was driven into the saw cut, behind the saw. This was on the opposite side of where the tree was notched. If a tree is nearly vertically straight, with the use of a wedge the tree could be made to fall any direction. Finally, I don't want to overlook the outhouse, but enough has been written about them so all I want to add is "thank God for indoor plumbing." One building that would not be found on the farm is a garage, at least not until after 1910 when the Model T began to be produced. Even after that, farmers would park their cars in their machine shed or some other vacant building. These buildings were without heat and most cars were reluctant to start in the winter. Even it they did start, they usually couldn't go far due to snow covered or muddy roads.

A barn raising, also historically called a barn rearing or bee, is a community action for a member of such community. This practice continues in some Old Order Mennonite or Amish communities. In pioneer days, communities raised barns as many laborers were needed and carpenters or barn builders as we know today were simply not available. These barn raisings provided a social gathering for the neighborhood when the whole family would get involved. While the men were working, the women would have either brought prepared food or be preparing food to provide a festive meal and possibly followed by a dance on the floor of the newly built barn. Barn lofts would be mostly empty of hay in the spring so they would very often be a place for a dance, a wedding shower, or various social gatherings, including box socials. Box socials were held to raise money for some need in the community, maybe to erect a new barn for someone whose barn had been destroyed by fire or wind or maybe to raise money for a country school, or any number of community needs. For a box social, young, single girls would make up a basket of several of her favorite foods, decorated attractively, and hope that a particular young man would bid on her basket so the two of them could share it together. Of course, no one was to know who brought the baskets, but most often the girls would drop a hint to the young man she wanted to share it with, such as what color bow might be on it or the color of the table cloth sticking out the top of the basket. I recall my mom telling about box socials and how she would make sure she had dropped a hint to my dad. The practice of box socials seemed to have lost its favor with young folks between 1970 and 1980.

Growing up on a farm, I have many fond memories of playing games with my siblings in the barn, especially the hayloft, whether it would be empty, partly full or all the way full of hay, it certainly made a fun playpen. We would roller skate on the empty floor, or swing from ropes and drop into the hay piles. We were quite creative about finding ways to have fun—sometimes not the safest way to play.

There is one feature of barns that I am reluctant to write about. Haylofts in early American and European barns were generally only accessible from the inside of the barn through what is called a trapdoor, about 4x6 feet, via a ladder. If the door was left open, it could present a danger factor as a person, especially a youngster, could easily fall down the opening and there are many stories where that happened. If the person was lucky, he would fall onto a heap of hay unhurt but less lucky and he'd probably end up with at least a broken arm. Most barns had a silo attached to one end, usually a shade higher than the barn.

There is an old expression, "Were you born in a barn?" This term originates from farmers letting their barn doors open so their livestock could go in and out at will for giving birth, depending on the weather. I'm sure some of my grandchildren recall being asked that question on many occasions as they, like most kids, run in and out often leaving doors open behind them.

Chicken houses and smokehouses had two things in common. First, they were located close to the family residence, and second, they both had to have locks on their doors which were kept locked, especially at night. This was the case even when the house the family lived in often had no locks on the doors. It was especially important during the depression when many folks simply didn't have enough food or money to buy food. Stealing chickens and smoked pork from their respective houses has been an age-old problem. During the height of the chicken stealing, the University of Illinois came up with various ways for farmers to prevent theft. One was, of course, to lock the sheds and another was to put a tattoo on the breast of the chickens. Often times there was another small building located close to the house which served as the brooder house or it may have been a small room attached to the chicken house. When electricity came to the farms, brooder hatcheries were started up in rural areas for hatching baby chicks. Farmers would buy day-old baby chicks from the hatchery that would then have to be kept toasty warm for at least two months, requiring an electric brooder, thus the brooder house. In grampo's day and many centuries before, without electricity, chickens and all other poultry were hatched by the hens sitting on their eggs (brooding) to keep the eggs warm until the chicks popped out of the eggs. The broody hens, also known as hatchy hens or clucky hens, took their motherhood seriously by faithfully sitting on their eggs using their body heat to keep the eggs at 100 degrees and turning them every day for 21 days. After the eggs were hatched the mother hen would gather her chicks together under her wings and sit on them to keep the chicks warm in cool weather. Which, by the way, caused me to think when I was very young that hens had a distinct advantage over other mothers. I wondered if God was fair when hens didn't have to give birth but most other animals had to go through painful birth. Well, I guess I was too young to even be thinking logically.

Today, most eggs are hatched in incubators, but if you decide to hatch your own chicks, I offer some helpful suggestions. Of course, you will need fertile eggs requiring a rooster to be included in your flock. You had better be very careful when you stick your hand under the cluck to check her eggs as she is quick to peck at your hand in her dedicated effort to defend her eggs until those cute little fuzz balls break out. More about chickens in Chapter 12, Part 6.

On our family farm when I was growing up, tame geese and ducks along with chickens would all contribute to filling up the barnyard. Male ducks are called drakes while the females are called ducks and their offspring are called ducklings. Male geese are called ganders, females are geese, and their babies are called goslings. The gestation period for ducks is near a month and a few days longer for geese. In the spring of the year, it was quite common for more chicken hens to get clucky than were needed, but they weren't relieved of their hatching obligations as farmers would put them in a box and put duck or goose eggs under them. Apparently, the chicken was more apt to take her mothering obligations more seriously than the ducks and geese. At the end of the gestation period, the babies would start to peck their way out of the eggs and within a day or two they would be out and about eating whatever was available. Their webbed feet would get a workout on the water, sometimes the same day they were hatched. It was quite the sight to see ducklings or goslings following their surrogate clucky chicken hen all around the barnyard, however their mom remained very motherly. Many times, I would see a cluck standing on the edge of a stream while her goslings or ducklings were having the time of their lives swimming and ducking their heads and occasionally making a dive into the water to nab a minnow or anything else that looked good to eat. The first times the cluck would take her little family of ducks or goslings to the stream, she would be pacing the bank, clucking away, trying to call them back to safety. She knew she couldn't swim and thought her babies couldn't as well, but after a while she would get accustomed to their escapades and when they tired, she would take them back home. It was interesting to note that if more than one hen took her flock to the water that when they came out of the water, they would go with either of the hens there, not necessarily the one that brought them. As a result, one hen may have 20 ducklings or goslings following her while another hen may have only 3 or 4. The babies didn't seem to care who their mommy was as long as they could claim one.

I happened upon a recipe from among a collection of my wife's grandmother for worm treatment for poultry. It reads, "Give Epsom Salt at rate of 1 pound to each 100 hens. Then place hens on ration of dry mash containing 2 percent tobacco dust. 100 lbs. ground corn, 100 lbs. ground oats, 100 lbs. middlings, 100 lbs. bran, 8 lbs. tobacco dust. Feed tobacco treatment 3 weeks. Discontinue for 2 weeks and then feed for 3 weeks more." It doesn't say if it works or that it is a guaranteed fix.

Hog houses always had a cement floor. This was quite essential as hogs had a nose that was especially designed for rooting. Adjacent to the hog house, usually on the south side, was an enclosed hog floor where hogs were fed corn and slopped. The slop trough was made of two-inch lumber. It would be 1½ feet wide by 1 foot deep and long enough to accommodate the number of hogs being slopped. Across the top would be a 1" x 2" board nailed to each side about 1 foot apart to keep the hogs from laying in the trough to keep cool. Another version of a slop trough was a vee shape where two 2 x 12 foot boards would be nailed together to form a "V" for holding the slop with a board nailed on each end to close the ends of the trough and serve as legs for the trough and like the square trough, a 1 x 2 inch board nailed across the top. This was quite essential as hogs are very sensitive to hot and cold weather. Hogs have at least one thing in common with humans in that we both like our surrounding temperature at about 70 degrees. For the convenience of slopping hogs, one end of the trough extended about one foot through the fence which enabled the farmer to pour slop into it without going into the pen and getting mauled. Slop was a delicious dessert for hogs, enjoying it even more than corn. It was from observing hogs go after their slop that prompted the saying, "Making a hog of yourself." At slop time, it was every hog for him/herself, pushing and shoving for their best advantage.

What were the ingredients of slop? Primarily, farmers sell milk as grade A milk, that is bottled for drinking, or sold as grade B milk which is made into cheese. The primary difference is that grade A has to meet higher sanitary conditions. Some farmers elect to sell grade A milk as it pays a premium over grade B, but it requires stiffer sanitary regulations and frequent inspections of the dairy barn.

In grampo's day, he and many farmers would have their own separators to skim the cream from the milk. The skimmed milk would be used to make slop for the hogs. The cream would be churned into bright, yellow butter that would be stored in the ice box (before electricity) and then taken to the local Saturday produce market or peddled to customers. Cream was also used to make home-made ice cream. The wooden ice cream making machines were very common in most farm homes and those who ate the ice cream from those machines will tell you there is no better ice cream made.

The grade B milk that was taken to the creamery would be separated into curds and whey and the curds would be made into cheese. The whey was all hauled back to the farm in cans where it would be poured into a barrel, add a few scoops of ground oats and table scraps, and used for slop for the hogs.

Curds and whey bring up the very old nursery rhyme from about 400 years ago.

"Little Miss Muffet sat on a tuffet,
eating her curds and whey.
Along came a spider
who sat down beside her
and frightened Miss Muffet away."

Little children can rattle off this rhyme but if you are to ask them what curds and whey are, you'll likely get a blank stare.

Many farmers had their own cream separators so they would haul the cream to the creamery and keep the by-product for making slop. My gram would keep a slop bucket close to the kitchen sink where all edible garbage would be collected and added to the milk by-product, along with a few scoops of ground oats, making for a good quality slop for the hogs. Without fully realizing it and out of necessity, our ancestors were good stewards of the land.

There is evidence that in Egypt cheesemaking goes back 4,000 years. By the time of the Roman Empire, cheesemaking took place in all parts of Europe and eventually all around the world. As time went on, many different varieties were developed.

As cheesemaking spread across the U.S., it was a farm enterprise until 1851 when the first cheese factory was built in New York State by Jessy Williams. By 1880 there were nearly 4,000 factories across the U.S. Farmers would haul their milk to city plants in two-wheel carts, wagons, or whatever means they had until the 1920s when trucks took over the task.

Recently I asked Clinton White to share his experiences of hauling milk to the Menominee, IL, cheese factory in the 1950s. He said his truck held 50 cans of which 45 were used for milk and 5 were used for whey. His route consisted of 11 farmers that sold from one to five cans a day. Two farmers had only one can. The milk in the ten-gallon cans weighed 80 lbs. plus the weight of the can would bring the total weight to nearly 100 lbs.—quite a load to hoist onto the 3½ ft. plus truck bed. I asked if anyone rode along to help. He said occasionally Elaine, his wife, would ride along primarily to open gates going into many of the farms. I then asked if he had any unusual experiences to share. He told of a bachelor farmer whose house burned down and put a roof over the basement to live in. One day when Clinton went to pick up his milk, he sensed something wrong as the cows were waiting to be milked. Clinton drove to the nearest neighbor for help. Upon returning, they broke in through a window and found the bachelor dead in bed. This bachelor sold one can of milk a day and milked his gentle cows outside by hand.

Clinton kept his milk hauling career for five years.

Grampo had a unique way of managing his Shorthorn cows. He would plan for them to have calves in late May. The cows would then nurse their babies through the summer months until mid-September. He would then wean the babies and milk the cows, separating the cream from the milk. The cream would be sold and the milk would be used to "slop" (feed) the pigs. His strategy was to avoid milking the cows in the hot summer months and to free up more time during that busy season.

In 1910 grampo built a new horse barn to accommodate his growing number of Persian horses. As I mentioned earlier, grampo's farm had a larger than normal amount of pasture acres. He chose to utilize those acres to expand his horse enterprise rather than more cattle. Even though grampo's ancestors were German, he chose to raise Persian horses that came from Western France. The exact origins have been lost over the centuries. Most Persians are black or gray in color and noted to be very docile. They were originally used as war horses and later as draft horses. Grampo kept up to 18 mares. His income from horses was twofold, i.e., selling the offspring and using his two stallions to stand at stud. He provided stud service either by having the neighbors bring their mares to his farm or for grampo to take the stallion to the neighbor to service his mares. Grampo kept the best quality stallions in the area so they were in big demand to improve the mares' shortcomings.

Before the invention of the combustion engine little more than 100 years ago, horses were the chief source of power for thousands of years. They provided the fastest and the surest way to travel. Hunters and sportsmen made good use of them as well as did soldiers. Entertainment should not be overlooked as quarter horses have been noted to run at 47.5 miles per hour, providing entertainment at the racetracks.

When grampo purchased his farm many of the buildings, including the house, were in a dilapidated condition. Building new buildings was the best option, which is what he did over several years to avoid causing too much damage to his wallet at one time. The horse barn was the first new building he built as it was the most important one. Sheltering his horses was the first priority of a farmer since without horses a farmer would be completely handicapped.

When our forefathers came to America, chestnut trees were estimated to be in the billions, native to Eastern North America. The chestnut tree could grow up to 90 feet tall and a diameter of up to 9 feet. These trees were comparable to other hardwoods such as oak, but grew faster and larger. As a hardwood, they had unlimited uses. The lumber was used for houses, furniture, split-rail fences (important before barbed wire was invented), piers, telephone poles, and for posts. They also had a very predominant use as flooring for horse stables since they would be more comfortable for horses to stand on as they got their night's sleep. Horses sleep standing up so the chestnut flooring was the chosen lumber for stables and also easier to clean than cement. The chestnut was considered to be one of the finest trees in the U.S. until a blight was found in 1904 which all but wiped them out in the first part of the 20th century. Horse owners now had to use other hardwood lumber to assure their steeds a good night's rest while hogs continued to sleep easy on their cement floors.

About four years after building the horse barn, a cow barn was built. Many buildings were built to accommodate the horses, cows, sheep, and other farm animals. Grampo wanted a barn for his cows and a separate one for his horses.

Last but not least is the woodshed, usually located to the rear of the house. In the fall, it was stacked full of wood providing the snow was polite enough to wait for farmers to fill it. Otherwise, cutting wood could have been an all-winter activity. The woodshed always had more than enough mice and trapping them out of existence seemed to be a hopeless case. The woodshed, or behind the woodshed was a very notorious place for corporal punishment. It provided a secluded place away from the rest of the family, and inside the punisher could find the appropriate stick to tan the punishee on the body part that would make it hard to sit on anything but a soft pillow afterward. Thus, the phrase, "I'll take you out behind the woodshed."

Chapter 8

Marketing Livestock

During grampo's farming days, the majority of the day-to-day labor was accomplished by the family members. There were some tasks that were community projects such as threshing grain, silo filling, butchering hogs, and shredding corn, just to name a few. Selling livestock could be done by the individual farmer or several farmers, especially if there were many head of cattle to be driven to the nearest stockyards, a distance of up to 12 miles.

Christmas Day in 1865 marked the beginning of modern livestock butchering with the opening of the Chicago Union Stockyards. They were soon

Buyers bidding on fat cattle for packing plants

53

receiving livestock from all areas of the country. There were 2,300 separate pens on 375 acres with pens that could accommodate 75,000 hogs, 21,000 heads of cattle, and 22,000 sheep, plus the yards contained numerous corn cribs and hay barns. Surrounding the stockyards were restaurants, hotels, saloons, and office buildings making up the world's largest meat packing district on a total of 475 acres.

Adjacent to the west of the yards, several meat packing plants were established, taking up a one-half square mile, known as Packing Town. The poet Carl Sandberg referred to Packing Town as the hog butcher for the world. A common statement made about butchering hogs was (and probably still is) that all parts of a hog were used except the squeal.

Selling livestock in Chicago was the best option for Midwest farmers and western ranchers as it provided a competitive and reliable market—very competitive with the several packing plants including giants such as Armour, Swift, and Wilson plants. These plants in Packing Town were the beginning of the use of assembly line techniques (or maybe "disassembly" line techniques fits better). These lines were the opposite of Henry Ford's assembly lines as the packing plants took livestock apart whereas Ford's put auto parts together.

The first practical refrigerator railroad car was not built until 1870, five years after the opening of the Chicago Stockyards. However, every Monday and Wednesday at 10:00 A.M., with a sufficient amount of wooden livestock cars, the Illinois Central Railroad would stop at the East Dubuque, IL, stockyards to load cattle, sheep or hogs, and occasionally goats (more about goats later) and be on its way to Chicago Union Stockyards, making numerous stops along the way to load more livestock.

Ben Kuhl was the manager of the East Dubuque stockyards located across and adjacent to the railroad tracks. He was also manager of the feed store located at the end of Main Street next to Jack's Tavern, just one of 26 places in East Dubuque where adults could purchase adult beverages. Eventually Ben's son-in-law, Phil Runde, took over the management of the stock yards.

When selling, farmers would load their hogs, sheep, and calves onto their wagons. If they had one or just a few cows to sell, they would tie them behind their wagon and lead them. Arriving in East Dubuque, the farmer would have to stop at the feed store to pick up Ben or Phil or possibly next door at Jack's Tavern where they might be sitting in as bartender and then on to the stockyards. If the farmer had many cows to sell, he would drive them on foot rather than load them onto a wagon. After unloading their livestock, the next stop would be at the watering trough by the police station as by now the horses were badly in need of a drink. As you might guess, the farmers were also in need of a drink or two. If they over-indulged, driving home was not an issue as all they had to do was untie the horses from the hitching post, hop in the wagon, steer the horses in the right direction toward home and the horses knew the way. (Automatic steering in the horse and buggy days.)

East Dubuque, with its 26 taverns and night spots, could take care of the farmers' thirst but here I want to detour to spotlight some highlights of East Dubuque, a town with a population of 1,671 residents according to the 2014 census. Originally the town was named Dunleith with a Post Office being established in 1854. The name was changed to East Dubuque in 1877. The first block and a half of eastern Sinsinawa Ave. beginning with the junction of Wisconsin Ave. had business stores for farmers and urbanites alike. There were two banks (both of which survived the depression), three grocery stores, a drug store, a feed store, Post Office, gas station, hardware store, and Grassel's Shoe Store to name a few. This first stretch of Sinsinawa Ave. had a few taverns, however as one drove westward on its street wide enough for diagonal parking of vehicles, taverns, night clubs, and gambling halls became much more numerous, practically the only buildings on either side of the wide avenue. An exception was Runde's Chevrolet and garage, at the far west of town, formerly a livery stable. When prohibition started in 1919, whiskey stills popped up around the area supplying those in need on Sinsinawa Ave., also known as Little Las Vegas, The Neon Valley, and The Strip. I have heard it said that during prohibition more whiskey was consumed in East Dubuque per capita than before or after. On the strip, of course, there were some things going on that many of the residents of the town did not approve of and East Dubuque became known as Sin City.

As East Dubuque is my hometown, I am compelled to defend a better side of it than its reputation. I believe it was the out-of-town visitors especially from Dubuque, Iowa who would come to East Dubuque after places closed in Dubuque to finish off the night. Some dignitaries such as Al Capone also contributed to some of the bad press for East Dubuque. The local folks, for the most part, were hard-working, ordinary good citizens and family folks who practiced their religion attending church services regularly at one of the two churches in town, Wesley United Methodist or St. Mary Catholic with its Catholic grade school. The parishioners of St. Mary's Church are quite proud of the unusual large numbers of religious vocations from their parish including eight priests, one bishop, and twenty religious sisters. I'm sure the Methodist church has its boasting rights as well. When I was a young lad, I knew nothing of East Dubuque's reputation as it wasn't ordinary kitchen table conversation at our home. My family only patronized the business district, and of course, the church.

Back then, it was the norm to wear hand-me-down clothes if they had any life in them at all. Shoes were repaired if at all repairable. It was at Grassel's Shoe Store where I got my first education in business dealings. One day on a shopping trip with my mom and brothers, Richard and Ned, we all had a task.

Mom went to Blums for groceries, while Richard went to Runde's feed store to get a bag of mash for the baby chicks. He also had to go to Jack's Tavern to get a can of plowboy Chewing Tobacco. Meanwhile, Ned and I were sent to Grassel's Shoe Store to pick up shoes that he had repaired. I would have been about six or seven years old, Ned was about four years older. Ned did not pay for the shoes which I noticed but didn't understand the reason. Mom had a charge account at the three grocery stores, Grassel's, and likely some others. I wasn't in the know about charge accounts. I concluded Harry Grassel fixed the shoes for nothing. With that thought in mind, about a month later, the soles of my shoes needed repair so I went to Grassel's during the noon recess and asked Harry if he could fix my shoes for nothing. His sarcastic response was, "What do you think this is, a play house?" Anyway, he sewed up my shoes. When I told Mom about this, she reprimanded me and said she would pay Harry when she settled her account. Whether he put it on the account or if Mom ever paid him, I'll never know, but I did learn about charge accounts.

Harry Grassel and his son, Fay, ran the shoe store for many years, selling Red Wing Shoes among others. They were both very hospitable and charming conversationalists, unless you wanted shoes fixed for nothing.

Now back to the marketing of livestock. Upon arriving at the Union Stockyards, the livestock would be sold to a packing plant buyer who paid the top dollar. The farmer would be mailed a check with any expenses such as freight etc., deducted. Yes, about the goats—once my dad shipped a goat and after the goat was sold, Dad received a statement from the Illinois Central Railroad for additional freight as the sale of the goat didn't cover the cost of shipping. Dad was obviously unhappy but reluctantly paid the bill. I was and still am unable to find out the disputed amount but apparently goats weren't the most profitable commodity.

The mutton from older ewes (ewes no longer suitable for breeding) is one of the less desirable meats so old ewes sold at depressed prices. Mutton also got a lot of bad press as it was occasionally used in rations for GI's especially during WWII. The same year that my dad sold the goat at a loss, another farmer sold a load of old, non-productive ewes that also failed to sell for enough money to pay the freight from East Dubuque to Chicago. The railroad sent the farmer a statement showing he owed freight money. The sheep owner responded, saying, "I don't have any money but I have more sheep."

For about a century or as long as the Chicago Union Stockyards were open, they were considered one of the city's world wonders. Visited by tourists from around the world, these folks marveled at the sight of the stock pens stretching for as far as the eye could see.

In grampo's time family vacations usually consisted of attending county or state fairs, hometown parades such as on the 4th of July, and various local events, rather than going on long journeys. Sometimes the family would hop onto a passenger train and head for the Chicago stockyards to see their cattle sold, tour the yards and a packing plant, and even possibly the wives might be able to take in some shopping at Marshall Field's department store or whatever favorite store. Marshall Fields was the finest department store in the land, stocking merchandise from not only the U.S. but also cities in Europe and especially the latest fashions from Pairs. I visited Marshall Fields about two weeks before Christmas in 1955 at about 10:00 A.M. The place was already overcrowded and by 11:00 A.M. there was absolutely no more room for more bodies. The entrance doors were locked for the rest of the day.

The Chicago Union Stockyards closed down on July 30, 1971. The stockyards in East Dubuque remained open as a buying station for the Dubuque, Iowa packing plant until 1951. Phil Runde continued as manager and buyer of livestock for the Dubuque Packing Co. until the East Dubuque yards were sold in 1951 and torn down the following year.

In the mid to late 19th century and up until the 1950s and 1960s, farmers, especially grain farmers, in the corn belt would purchase up to a train car load of calves from western ranchers to feed during the winter and into the following summer when they would be finished or ready for slaughter. This practice continues today but on a much larger scale as huge feedlots began to expand in the 1950s and '60s in states like Texas, Oklahoma, and Nebraska. This plan for the small cattle feeder had some good and bad results. The good was that the farmer could sell his corn through cattle at a premium price, meaning the corn would be fed to the cattle and the cattle would be sold. It also provided him with much-needed fertilizer, especially before commercial fertilizer. Another asset was that the farmer or hired help's time would be better utilized. One of the downsides to feeding cattle was sickness that could run up veterinary bills and even death losses. Sickness in livestock was a bigger problem in grampo's day than today. In recent years veterinary medicine has come a long way in protecting livestock and saving lives. Another problem was the volatility of the market. The price received at the time of selling might be so much lower than the purchased price, in spite of the weight gain, resulting in the farmer losing considerable money in the process. Most often the farmer would have taken out a loan to purchase the calves and if things didn't go the way planned, the farmer could owe money on the loan after selling all of his fattened cattle.

The hilly and rolling country of Northwest Illinois proved very suitable for beef cattle. Lester, a close neighbor of mine, had a farm that fit that description. He raised and fattened calves produced from Black Angus cows. When his cattle were fat and ready for slaughter, they would end up at the Chicago Union Stockyards. Lester made it a yearly event to go to Chicago to see his cattle sold. I recall him telling me a sad experience he witnessed one time when he was at the Chicago stockyards which expresses the difficulties many farmers ran into in marketing their livestock.

Early in the morning buyers began to make the rounds looking over and buying the cattle they wanted for the day's kill. On this occasion which Lester tells about, a buyer offered a farmer from Western Iowa a set bid for his cattle. The farmer refused and tried to negotiate for a few bucks more to no avail. Other buyers came along but would not offer any more than the first buyer. Of course, the farmer held out hoping for a better price but as the day went on the market prices went south. The farmer had to sell for considerably less than the original offer resulting in the farmer losing money. Lester continued to tell me how the farmer began "to cry like a baby" as he explained that he wouldn't have enough money to pay the original loan and worried that his banker might foreclose on him. I'm certain this was not an isolated case. Lester's 175-acre farm was a short distance from his main farm having no buildings. It had a river running through it with steep hills and bluffs on either side. It was mostly pasture and timberland with some tillable bottomland that produced a bumper crop of corn if flooding didn't occur. Lester's farm was a land of Eden for all species of wildlife, such as beaver, muscrats, wild turkeys, opossums, mink, deer, turtles, squirrels, skunks, groundhogs, rabbits, foxes, an occasional snake, and many raccoons. It was also notorious for producing a bumper crop of mosquitoes and insects of all species—enough to keep an insectologist busy for a coon's age. On top of one of the hills was a small field, probably large enough for modern machinery used today such as a 32-row corn planter to turn around on. There were also an above average amount of fences and floodgates that provided entertainment for Lester and his son, John, especially on rainy days. To conclude, this farm probably had more value as a happy hunting ground than for agriculture. Lester, a devout hunter and fisherman, always got his allotted number of deer in spite of his unique way of aiming and firing a rifle. Instead of holding the butt of the rifle against his shoulder, he would hold it against his chest, somehow aim and squeeze the trigger, usually dropping a deer. The accuracy of his aim still puzzles me.

Another marketing story I have heard told many times is one I'd like to share which is on a much lighter side than the previous story. This little piggy was actually a big piggy, a boar that weighed close to 550 lbs. Reuben owned a farm on the northeast part of East Dubuque. His farm was by no means an all-tillable farm. Some folks referred to it as Billy Goat Mountain. In Reuben's day, pickup trucks gradually began replacing the horse and wagon to transport livestock to market. Starting in 1910, one would observe a mixture of horse-drawn vehicles and cars. By the mid 1920s, dusty pickup farm trucks came on the scene. Dirt roads, like most if not all rural roads, caused pickups to become dusty enough for notes to be written on the dashboards. It was also quite common to see these early motorized vehicles stopped in the streets or along the roads, while the horse-drawn vehicles whizzed by. Model Ts were built from 1908 to 1927. These early vehicles sometimes had their unique personality. They didn't like to start and before the invention of electric starters, they had to be cranked to start. The cranking was good at breaking one's arm as the engine would kick back, causing an outburst of profanity from the one cranking. Back to my story: When Reuben went to load this "little piggy" is when his troubles began. He backed up his pickup truck to the make-shift loading chute and began to drive little piggy up the chute into the truck. Little piggy would go to the top of the chute and stop. He wanted no part of going into Reuben's pickup. Do you suppose he had a premonition as to what was to come? Some may dispute me, but I don't believe that was the case. Anyway, Reuben tried several times and each time he drove little piggy up the chute, little piggy refused to go into the pickup. Reuben finally got disgusted. He tied a make-shift halter on little piggy, tied one end of a lead rope to the halter and the other end to his pickup. Reuben hopped into his pickup and off they went as they headed through a residential area, then the business district, and off to the stockyards. Little piggy protested all the way squealing as loud as a pig can squeal on this 20-minute trip to market. I assume East Dubuque didn't have a noise ordinance at that time. At any rate Reuben and little piggy provided a few minutes of entertainment for the town folks who had gathered to see the noisy mini parade. I can't say for sure, but I would guess that story made front page news in the weekly *East Dubuque Register,* as few local happenings escaped that newspaper, like most small-town newspapers, including world news and church and community events. However, I regret that I was unable to find a picture of Reuben towing his pig to market. People still remark about reading about who poured the coffee at a private party and what ladies attended the latest sock darning get together. Throughout history farmers took livestock to market in unusual ways. I recall seeing in a city in Brazil, a farmer pedaling a 400 pound sow to market in a wheelbarrow providing entertainment for our group of tourists. At least that pig wasn't squealing, but seemed quite comfortable and contented riding along in the wheelbarrow. I do have a picture of that scene and it can be found with other pictures in the center of this book.

During the Great Depression of the 1930s, farmers had to be ingenious in order to survive. I heard of a farmer with a few head of hogs, having no way to get them to market because it was too far to drive them on foot and he had no truck. So, his ingenious way was to put two at a time in the back seat of his car to transport them which worked as long as his gas-rationing stamps didn't run out. For farmers who lived close to a city's packing plant or stockyards, it was not uncommon to see a heard of hogs paraded through main street on their way to market.

Bartering is another system for marketing produce, but without having money involved. It has been around for centuries before money appeared and to some extent is still used today. In my mom and dad's time, chickens were one of the prime items for bartering. During the depression, my dad would barter chickens, eggs, and potatoes for labor. He would also barter two chickens every Saturday to pay the dentist. It was the norm for mom and dad to start early on Saturday mornings, pluck six chickens, two for the dentist and take four to the Farmer's Market along with other produce to be sold for cash. Other poultry such as geese, ducks, and turkeys were marketed directly to consumers, mainly our city cousins, especially during Thanksgiving and Christmas.

Ann's mom would barter eggs at the local grocery store for groceries. She would take a case of 36 dozen eggs to town once a week along with her grocery list. The grocer would subtract the price for the eggs from the cost of the groceries and put the balance on her charge account. Maybe sometimes the amount for the groceries would be deducted from the amount for the eggs and there just might be a deduction on the charge account. At the end of the month her mom would pay the remaining bill.

Loading hogs on a semi to sell. Today's hogs ride to market in a comfortable semi.

Chapter 9

Livestock Sickness and Treatments

Concerns for diseases no doubt dates back to man's first taming of wild animals. Early treatments consisted of herbs, religious practices, and magic. It is also known that Egyptians would apply moldy bread to sores. A few thousand years later, moldy bread led to the discovery of penicillin. (More evidence that we can learn from our "mummies.")

From archaeological findings, it is believed that people were treating animals in 3000 B.C. Throughout history, horses got the best care and treatment of all animals as they were the most valuable animals, at least until the 1940s when they were replaced by combustion engines in many ways. A serious cattle plague broke out in France in the mid-1700s. This triggered the opening of the first veterinary school in Lyons, France, in 1762. Cattle plagues were a notorious problem as long as can be remembered.

It wasn't until 1852 that a veterinary school was opened in the United States in Philadelphia. The following are a small sample of treatments for various diseases from a book loaned to me from Dr. Cal Schafer, DVM. The book's title is, *Gleason's*

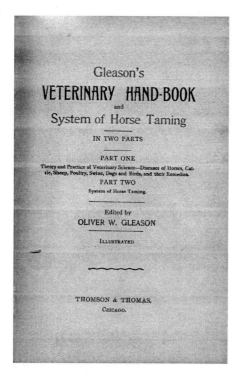

62 Changes In the Good Life

 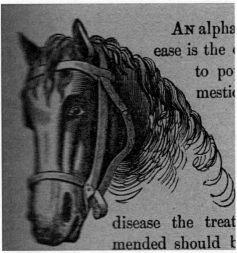

VETERINARY HAND-BOOK and System of Horse Taming published by Thomson & Thomas in Chicago in 1889.

JUNIPER BERRIES—Juniperis Communis—*The Fruit. This medicine is valuable in horse and cattle diseases, as a stimulant to the stomach in loss of appetite, and in convalescence from debilitating diseases.*

OAK BARK—Quercus Cortex. *This is a good astringent for outward use, or for sores which discharge or run matter. The bark is boiled: half an ounce to a pint of water. This decoction is an excellent remedy for drying up the moisture of greasy heels, so troublesome in horses. In diarrhea in calves, given in four drachm doses, much good will result.*

CREOSOTE —*This is a peculiar smelling fluid derived from tar. Creosote has had the credit of curing glanders in man, and is good remedy in pleuro-pneumonia in cattle, but we have better ones, and not so costly. Cases of arcy and glanders in the horse are greatly benefited by its use.*

CROTON OIL—*A dangerous medicine when improperly used, but a useful one nevertheless, when hasty action of the bowels is wanted, as in milk fever in cows. Dose: For the cow ten to fifteen drops, given along with Epsom or Glauber salts.*

For the dog, castor oil may be a proper and useful purgative; and for the pig, also. Aloes, and linseed oil, is the purgative for the horse;

epsom or Glauber salts for the ox, and the sheep. Whatever suites man, as a purge, will answer for the dog and pig.

CHARCOAL—Occasionally given to cows, in chronic diarrhea. Dose: Half an ounce to one ounce, given suspended in gruel, of any kind. Externally, charcoal is very valuable, when applied to badly smelling wounds, and ulcers.

ELM BARK—Slippery. This bark, when scalded with hot water, makes a useful poultice for irritable wounds, ulcers, and sores. A decoction of the bark will answer every purpose for which flaxseed, or linseed is used or recommended, as in diseases of the kidneys, and bladder, produced by the use of Spanish fly, and from over-dosing with rosin and other diuretics. In diarrhea, in all animals, slippery elm teas, or decoction, will serve a good purpose, by sheathing the covering of the bowels, which is so apt to become irritated and inflamed in violent superpurgation.

SPIRITS OF WINE—Alcohol. This is used for making tinctures for medicinal purposes, from the various plants in use. It is also a good stimulant; much better than the bad whiskey which is so often poured down the throats of horses affected with colic.

SPIRITS OF NITROUS ETHER—Sweet spirits of nitre is well known to most persons as a good household remedy for fevers, etc. In the treatment of diseases of horses and cattle, sweet spirits of nitre is used as a stimulant and antispasmodic. It is also used in the case of a horse having a chill, and in colic. For colic, it was formerly given in combination with laudanum. Dose: For horses and cattle, the dose of sweet spirits nitre will be from one to two ounces, given in cold water to prevent loss.

ALCOHOL—Spirits of wine entirely free from water, and is used for making tinctures of the various plants. It is the foundation of many lotions and liniments. Alcohol may be given to horses having a chill, in half pint doses, mixed with a little warm water, not too hot.

WILLOW BARK—A much neglected, valuable and cheap medicine. This bark has within it a crystalline substance called "salicine," which is an excellent substitute for the expensive quinine. Farmers and others will do well to gather it in sufficient quantity, and have it dried; and in the spring of the year, or when any of the horses are weak, or out of sorts, take of the willow bark one pound, and boil in four quarts of water, till two quarts are left; then strain for use, and give a tumblerful, mixed in cut feed, once or twice in the day. This will be found much better than the black snakeroot already spoken of under its proper head.

TOBACCO—Nicotiana Tabacum. *Tobacco is used as a medicine, principally in skin diseases, and for the destruction of lice and other insects in the wool of sheep. Tobacco smoke is a favorite remedy with some veterinarians, for removal and killing of worms, and in constipation, and colic. For these purposes, better and safer agent are in every-day use. Tobacco in all, or any of its forms, is dangerous, being followed by great sickness, nausea and prostration, from which many animals are ultimately destroyed.*

CATCHING A PIG—*The following method of catching a pig has been recommended: Fasten a double cord to the end of a stick, and beneath the stick let there be a running noose; tie a piece of bread to the cord, and present it to the animal; and when he opens his mouth to seize the bait, catch the upper jaw in the noose, run it tight, and the animal is fast. Another method is to catch one foot in a running noose suspended from some place, so as to draw the imprisoned foot off the ground; or, to envelop the head of the animal in a cloth or sack.*

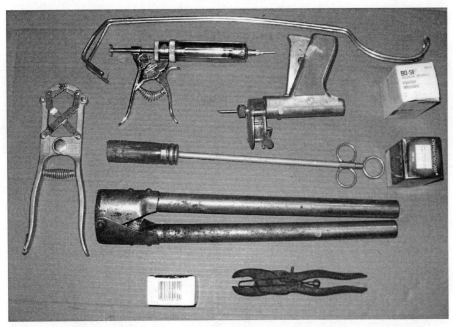

Tools used by veterinarians & farmers. Note the hog ringer which is rarely used today since confinement.

Livestock Sickness and Treatments

The control and treatment of diseases in livestock made sluggish progress until 1928 when Alexander Fleming accidentally discovered penicillin. After his discovery, progress continued more aggressively with vaccines and antibiotics and continued to change dramatically through the 20th Century.

When my grampo and my dad were farming, hog cholera, a highly contagious disease, was the most devastating and talked about disease in the U.S. with the first known outbreak occurring in Ohio. I say, "talked about" as I still remember stories told by farmers of past generations. One was about a farmer driving through Iowa seeing a common sight of smoke coming from farms where hogs, dead of cholera, were being burned. I was told how farms that had cholera were quarantined and the importance of sanitation. An outbreak of cholera could be a breaking situation for a farmer as the hog profits were most often used as the mortgage lifters.

On my dad's farm, for example, each commodity had its goal. The hogs had to pay the mortgage, chickens and eggs had to buy groceries, clothing and household items, profits from garden produce and wild blackberries sold at market were used to pay the real estate taxes. Sheep, wool, and the dairy cows had the biggest job of paying for machinery, building and fence repairs plus all other farm expenses.

I have heard stories of farmers losing their farms as a result of a cholera outbreak—no hogs—no money to pay the mortgage—no farm. My dad fenced off an isolated tract of land that had spring water. He built a temporary hog house there thinking that the virgin soil would be free of cholera. He also vaccinated with the questionable vaccine available at the time. At any rate, he escaped the dreaded disease. Whether the vaccine or the virgin soil was a deterrent was never known but he continued to make the mortgage payments.

In 1907 the first cholera vaccine was developed leaving scientists to continue research to find a more effective one. It took until 1978 to declare the U.S. free of cholera.

To talk somewhat knowledgeably about modern day veterinary practices, I consulted with Cal Schafer, DVM, to whom for many years, I delegated the job of keeping my livestock healthy. First, about Dr. Schafer. He was a native of Perry, Kansas, and graduated from Kansas State University in 1976. Upon graduation, he joined the Veterinary Associates of Galena, IL, and Hazel Green, WI. He moved to Galena rather than practicing in his home town of Perry because there were few livestock in that area. More importantly, he couldn't resist living in the rustic beauty of Northwest Illinois, especially in Jo Daviess Co., or at least that is what I'm supposing. I met with Dr. Schafer at the Galena Veterinary Clinic early one morning and started asking questions.

Bert: How do larger farms and consolidation of livestock affect veterinary practice?

Dr. Cal: In this area, there are still numerous small dairies and beef cow herds. The average beef cow herd in Illinois is 33 cows. There are many such herds in Jo Daviess Co., as our rugged terrain and pasture land is best suited for grazing beef cattle. We also care for horses owned by farmers and horse lovers.

Bert: How have veterinary practices changed since you first started in 1976?

Dr. Cal: There have been numerous changes. Today, we focus on herd medicine rather than individual animal medicine. One significant change is that today we are more concerned about producing healthy food. When I got out of vet school, we didn't look for residue in meat. The veterinarian's goal is to produce a healthy product. This requires teamwork between the livestock producer and us. It's our responsibility to raise a healthy animal that is safe for the consumer.

Bert: Do the lay people (producers) do some or much of their own animal treating?

Dr. Cal: Yes, many are trained to do this. We, here at the clinic, occasionally offer classes focusing on such things as giving shots in the right place, withdrawal of medicine, and much more.

Bert: With fewer and larger livestock operations, are there a sufficient number of vets or too few to care for large animals throughout rural America?

Dr. Cal: Not nearly enough. With less vets, it is more difficult to provide services needed. There are fewer coming out of vet schools that are interested in food animal medicine.

Bert: How have vaccines and medicines improved in recent years?

Dr. Cal: Vaccines today are much more effective for immunity. We emphasize immunity. An example would be scour-guard vaccines given to cows to avoid diarrhea in baby calves. However, if this practice is neglected, there is an effective medication for the disease also. Pneumonia in cattle is still a problem. Vaccines and treatments are available, however, they are expensive and not always 100% effective. It's also important to note that livestock drugs made today are designed to produce low drug residues and short withdrawal times.

Bert: What are issues or problems facing veterinarians today?

Dr. Cal: That is a subject I would like to address. I am glad you asked. The need for bio-security in America is hardly recognized and very lax. We veterinarians are the first line of security. It is quite easy for bio-terrorists to enter the U.S. and spread livestock diseases. I worry about this. This critically important issue is rarely talked about on the news as the news is too focused, in my estimation, on other less important issues. However, our Secretary of Agriculture, Sunny Purdue, is a veterinarian and he is showing concern about the issue. I would like to see a better identification system for cattle so they can be traced back to the original owner. So, if a steer is slaughtered in Nebraska and went through three sale barns, he could still be traced back to Bert Tranel's farm, his place of birth.

I thanked Dr. Cal for taking time from his schedule that day which would consist of a herd check and two horse calls. The herd check was at a farm where the farmer wants to do seasonal milking, avoiding milking during the winter months. Doc said it was his job to make sure the cows were bred at the right time so the lactation period would end in December and freshen again in early March. I then asked him just what is a herd check.

Dr. Cal: In recent years, doing a dairy herd check has become a common practice, done every 3–4 weeks in most herds. This represents an estrus cycle for the cow. While doing a herd check, we do a pregnancy check, note the condition of their udders, check for lameness and the overall health of each cow. The owner will usually have less problem cows for us at the time of the visit.

Chapter 10

Farm Sounds of the Past

While writing about farm changes, I don't want to overlook "sounds" that were heard in yesterdays gone by. Many farm sounds disappeared since my youth and it is not likely they will ever return. Some sounds like roosters crowing can still be heard on some farms and especially on isolated ranches. It is still common for a rancher to keep a small flock of chickens and a milk cow for a fresh supply of eggs and milk. In my youth, I recall hearing roosters crowing on our farm and most of the neighboring farms; not so today. I enjoyed hearing them. I guess there were some folks who found the sound very annoying. About 15 years ago, a neighbor quit raising chickens and I haven't heard a rooster crow since. I have read about some small cities passing ordinances allowing residences to keep five hens cooped up in their back yards but no roosters. I'm sure there would be many complaints in the neighborhood about roosters crowing in the dawning hours to awake adults and especially small children. Apparently, the noises that would be most annoying to me, such as Jake breaks, screeching tires, sirens, honking horns, train whistles, airplanes

Cowbell

Rooster

taking off or landing, and barking dogs, are better tolerated by others. I guess our tolerance for certain noises depends on what we associate with the noise or what we have become accustomed to. Either the noise elicits a pleasant memory, a comforting feeling, or is just an irritating annoyance.

The putt-putt sound of the John Deere two-cylinder tractor could be heard on farms starting in 1918 when they first came into use. These putt-putts were built until 1960, although several different models emerged during that time. Some people hated the sound of the John Deere tractor while others found them to signify power and stability. I am most familiar with the six-speed transmission model B, the first row-crop tractor that appeared on our home farm in the early 1940s. Although it didn't have an electric starter, a fly wheel was used to start it instead of a crank. When I got old enough to drive it, I thought this was a big improvement over cranking. I didn't have to worry about getting a broken arm from cranking. There are still putt-putt tractors around but with limited use for small jobs. Their greatest jobs are for use in parades pulling small wagons or trailers, etc.

The howling of wolves is another of those sounds that you like or hate. When my great grampo arrived in Menominee in the late 1860s, wolves had nearly all disappeared from Illinois. I do recall a story from Grampo John Kieffer telling how his dad, Nicholas, told of hearing the howling of wolves. What few were left were soon eradicated as they were farmers' most hated predator, preying on nearly every kind of livestock and poultry available. Extermination programs such as putting a bounty on them, and loss of habitat nearly made them extinct in the lower 48 United States. By 1973, the U.S. Fish and Wildlife Service put them on the list of endangered species. From 1995 to 1997, USFWS released 41 wolves into Yellowstone National Park. As you might guess, this set off a feud between the ranchers near the park and the USFWS. The animal lovers and animal rights folks also got involved and still are. No explanation is necessary why the ranchers opposed wolves. I hear wolves do leave the park and go to the nearest ranches now and then to feast on a young calf or lamb. My understanding of the law is that ranchers can't kill wolves unless they are attacking their livestock. I also heard that occasionally ranchers kill wolves that are trespassing on their ranches regardless of the law. The next story I heard was how to dispose of slaughtered wolves without getting caught. Their solution was to put them on a train loaded with coal heading east where they would end up in Chicago. If anyone is wanting to hear the howling of wolves, the best place in the U.S. is at Yellowstone National Park or a nearby ranch. However, there is an occasional sighting of a wolf every now and then in the Midwest.

Cowbells which are hung around a cow's neck would be one of those rare sights and sounds today. The original purpose of the cowbell was to scare off predators. Were they effective against wolves? I couldn't find an answer but

I would guess not or at least not for long. Another purpose for the bell was to assist the farmer, especially the dairyman, in finding his cows when the twice a day milking time came. In this area of Jo Daviess County or farms in areas with the same terrain, the bells helped the farmer locate his cows in large pastures consisting of timberland and hilly land. I would like to point out to the younger generation that ATV's were not around forever. Even though the name implies that cows were the only bearers of the bells, cowbells were also found around the necks of sheep, goats, and various other animals, even cats and provided a certain enchanting sound as they moved about the hills and valleys. Today with milking parlors and robotic milkers, dairy cows hardly go to a pasture, making a cowbell quite obsolete.

I still recall when I was little, my older brothers, Ralph and Richard, planting corn with a team of horses pulling a two-row corn planter. (Some of the following might be a repeat from the "Corn Chapter" but told a bit differently so I'll repeat it anyway.) The two-row corn planter had a long wire attached to one side of it stretching from one end of the field to the other, held there by an iron stake. The wire had knots on it every 42 inches apart. Before wire was invented, rope was used with a knot tied in it every 42 inches apart. Ropes go back to prehistoric times, although it wasn't until about 4000 B.C. when the Egyptians invented a tool for making rope. The mechanism that held the wire attached to the side of the planter was called a fork. As the horses would advance the planter, the wire would slide through the fork and the knots would hit a trip causing it to open and drop the desired number of kernels (usually 3 or 4) through the shoe to the ground.

The 42-inch spacing was the desired width of corn rows in horse-powered days as that was the width needed for a horse to walk through. Now, rows are planted close to 15 inches apart. When the team and planter reached the end of the row, the planter operator stepped on a lever that would disconnect the wire, then he would turn the team around, move the stake and wire closer to the planter, hook it up and tell the horses, "getty-up." The knots would stay in place as the planter moved back and forth across the field so that the trip would be activated at the same spot as the previous rows, making the rows perfectly parallel. This is called "checking." By this method, the corn rows could be cultivated longways, crossways or even diagonally. The clicking of the knots, the sound of the trip were pleasant sounds to farmers as they were sounds of progress to them. This method of corn planting was the norm until the 1950s when herbicides, of which 2-4-D was the first, came into use eliminating the need for cross cultivation.

Picking corn by hand was the norm until mechanical pickers were introduced in 1909, but it was not until the 1940s that the one and two-row pickers powered by tractor engines were becoming popular in this Midwest area. Picking by hand was one of the hardest jobs on the farm. To aid in this task, a farmer used a husking peg. This was a device worn on the hand with a pointed end to spear the husk at the tip of the ear pulling it toward the butt of the ear and then snapping the ear loose from the stalk, allowing the person picking to throw the ear into the wagon. The pointed end of the peg would be toward the thumb of the palm of one's hand and strapped with a leather strap around the hand.

Picking corn traditionally begins in the harvest month of October. October 10th is the average date for a killing frost in much of the corn belt. When corn was harvested by hand, frost had good and bad effects. It killed the weeds that the cultivator missed, which were numerous as this was before the use of herbicides. Frost also killed the corn stocks and leaves, aiding in the drying of the ear. Earcorn had to be 20% moisture to keep in the wooden corn cribs. Today, most farmers have a grain moisture tester but back then, the farmer would drop ears in a stock tank. If the ears floated, they were dry enough to crib. It was much easier to walk through the corn rows with the weeds dead and the dried corn stalks.

To pick 100 bushel of corn per day, a picker would have to start early in the day when frost was still present on the ear. If the frost melted, the double thumb glove would be a soggy mess. Picking after a rain or snow storm would cause wear and tear on the glove as well. The design of the double-thumb glove enabled it to be turned over when one side was worn out. General stores in the corn belt were well stocked with double-thumb gloves as they would disappear in a hurry during the picking season.

The wagon that the corn was thrown into was equipped with a bang board on one side. The sides of the wagon were about 3½ feet high with one side having boards extending four or five feet higher which was the bang board. This allowed the picker to throw the ears at this side and the ears would not fly over the wagon. Persons nearby could hear the constant sounds of ears hitting the bang board. I suppose the sounds I recall affected me more because it was generally less noisy around us in these times without the sounds of motors running and drowning out so many other sounds. Today, we are encouraged to wear earplugs to cut out the loud sounds around us. Therefore, we miss all the more natural and pleasant sounds around us.

The team of horses pulling the wagon were so trained that when the picker hollered "getty-up" they would go and when the picker hollered, "whoa" they stopped. My 94 year-old-brother, Richard, was the best corn picker I knew. He tells of how he once had a well-trained team of horses that would advance forward when they saw him approach the front of the wagon and then stop when he hollered, "whoa."

Horses, with their near 360-degree vision, could see where the picker was without turning their heads. They didn't even need a rearview mirror. They were blessed with large eyes and 360-degree vision from pre-historic times in order to keep a watch for predators, such as tigers or lions. If you have ever noticed two horses standing in a pasture, you will note they will be standing facing the opposite direction next to one another. This is a natural instinct to keep an eye out for danger. Another unique feature about horses is that they can sleep with their eyes open and they have a built-in mechanism, a stay apparatus, that keeps their legs locked in place allowing them to sleep standing up.

A picker could pick 100 bushel of corn per day and unload it. That would be two wagon loads. He had to start early in the morning to accomplish this. Often Richard would get started on his third wagon load before dark. He had much needed strong arms but he said it took a week of picking to get his arm muscles built up and during that time, his arms would be very sore. I've heard it said that a good picker could keep an ear of corn in the air at all times as he went down the rows picking and pitching the ears into the wagon.

To unload the wagon, the corn had to be shoveled with a scoop shovel into the corn crib. The wagon had an end-gate about four feet long that would be let down level with the floor of the wagon to assist in shoveling the corn off the wagon without avoiding spilling any corn on the ground.

Before hybrid seed appeared, a box was attached to the front of the wagon where the better ears were kept for seed for next year's planting. Occasionally, a red ear was found. We younger kids would crawl through the corn crib searching for all we could find. It was a common belief at our house that Santa's reindeer had a sweet tooth for red ears of corn. We wanted to stay on his good side. We'd usually find a gunny-sack full that we put by the gate of the picket fence around our yard, on Christmas eve. The next morning the sack of corn was gone. While other kids left cookies and milk for Santa, we seemed to be more concerned with feeding his reindeer. I guess that was our farmer's concern for livestock in us.

An alternative to picking the corn was to let the hogs into the cornfield. This was known as Hogging Down Corn. They did this with their spring crop of pigs. The norm in grampo's day and before his day, was to have two crops of pigs every year, spring and fall, when the weather was not so hot or not so cold. Baby pigs are very sensitive to extreme cold weather. If they should happen to be born on a cold night, it was a common practice for farmers to stay up late to care for the babies making sure they would get a suck of milk and not wander out of the nest. This could sometimes be an all-night job. The gestation period for sows is three months, three weeks and three days. Farmers would add to this period by saying, "3 months, 3 weeks, 3 days and 3 o'clock in the morning." Today with modern heated and cooled hog facilities, farmers arrange to have the sows give birth year round, winter to summer.

Fall would be the ideal time to let the spring crop of piggies into a controlled-portion of the cornfield. After about two months these little piggies would be big pigs and sent off to market. I mentioned a controlled area. This was necessary if they were allowed to graze on a large corn field, such as more than two acres, or they would make pigs out of themselves by knocking down corn stocks and not eating all of the ear. On our home farm, we had a one-and-one-half acre field by the hog house that was planted to corn for hogging down. Other farmers had similar small fields for that purpose.

Chapter 11

A Final Sale

Historically, most farmers do not want to own ALL the land; they just want to own the land that borders their farm. And most farmers think it would be great if their children, grandchildren, and beyond would desire to carry on the tradition of tiling the soil. Grampo was no different than most farmers in this regard.

From 1900 to the late 1920s, grampo was a progressive and prominent farmer. As I mentioned earlier, he built a new house, two new barns and a large granary, plus made numerous improvements. He was the first one in his neighborhood to have a sulky (riding) plow. He purchased additional land next by adding 40 acres to his original 283. When his oldest son, Hyacinth, was ready to strike out with his own land, grampo purchased 80 more acres or a half quarter section for him which was referred to as "the 80." Later on, he purchased for his son, Louis, 40 acres. During this period farming was good to grampo. Besides

Farm Auction Sale run by Mike Koehler.

his farm holdings, he had stock in the State Bank of East Dubuque. One thing he didn't own was a car and he never had a driver's license. He was at one time noted for being the wealthiest person in Menominee Township. During this period from 1900–1910, the average value of farmland increased over 100%. This kind of land increase promotes speculation by farmers and often hurts them in the long run. When the U.S. got involved in WWI, a continued, roaring farming economy caused land speculation to excel even faster. Especially during the 1920s, many farmers would borrow money to buy more land or machinery, pledging their existing assets as collateral. This went well until the stock market crash on Black Tuesday, Oct. 29, 1929. From 1929 to 1933 the unemployment rate jumped from 3.2% to 25%. Farmers had produce to sell but people didn't have money to buy it. The bottom fell out of their assets of land, livestock, and machinery and remained flat for the next years. This scenario soon left a lot of farmers with more liabilities than assets. Grampo was one of them. The banks got mighty nervous. In 1930, farm foreclosures numbered in the thousands as well as did bank foreclosures number in the thousands. As the Great Depression wound down, foreclosures became less and less until they became front page news again during the Farm Crisis of the 1980s.

The dictionary defines foreclosure as the action of taking possession of a mortgaged property when the mortgager fails to keep up their mortgage payments. When a bank elects to foreclose on a person owing them money, they need to go through the legal procedure of a sheriff's sale. The sheriff's department takes charge. Occasionally a penny sale would take place during the depression. A penny sale, according to my mom who lived through the depression, went as follows. Before a sheriff's sale got underway, the neighbors, relatives, or close friends of the delinquent farmer would get together and discuss how they would help this farmer regain all of his possessions at a fraction of its value. These neighbors would warn anyone planning to bid more than a few pennies on any item would suffer a severe consequence like the burning of their barn or other farm buildings before they returned home. My mom said she heard of one farmer who recovered all of his possessions for less than $5.00. The bank got stuck for the rest of the loan. It was also necessary for the neighbors to get the auctioneers to comply by making threats to them. Of course, their commission on a $5.00 sale would be mighty slim pickings.

On March 10, 1936, no such thing as a penny sale could be pulled off when the sheriff closed down on my Grampo Kieffer. From 1929 to 1936, he barely held on to his assets. Financially things kept getting worse and by 1936 there was no longer hope for him.

This, of course, was before my time. My mom told me that the day before the sale, most of his possessions were taken to the East Dubuque Stockyards. Most of his livestock and other properties were hauled in a truck but not his horses. When telling this story about how her dad's prized Persian horses were led to town, my mom got teary-eyed. She had a vivid vision of her dad's 18 plus horses being led on the township road that went through and along my mom and dad's farm. They were well-groomed and looked fit for a parade. They were led in columns of two. The two stallions led the way with their tails tied in knots and a lead rope tied to their tail and to the halter of the horse behind them. This sequence went from horse-to-horse. The first six horses were harnessed, as the harness had to be sold too, with leather harnesses decorated with ivory rings all over it and highly polished brass rivets and hardware shining brightly. Included in the parade were two colts and four fillies. While parading into the stockyards, the young fellows walked mighty close to their mommy's sides, especially as they got closer to town. They became frightened of the unfamiliar and mysterious sounds and sights like horns honking, train whistles, numerous humans walking amidst tall buildings. This was all strange territory since it was the first time they left the security and comfort of verdant pastures and their place of birth. This trip, no doubt, took a long time as the parade needed to stop periodically so the colts and fillies could get a quick snack from their mother. Mares are different from cows in regard to milk production as God created horses for work and cows for producing milk. Horses have a small udder with two nipples. They give a small amount of milk, although rich in fat. This requires their babies to suck often, every one to two hours. Thus, it was necessary to stop the parade occasionally so the youngsters could get a tummy full of nourishment. In contrast to baby calves, cows have large udders with four nipples. Baby calves can get by with three or four feedings per day. One other episode that held up the parade was when a three-month-old colt got too close to the side of the road by a culvert and fell in a ditch. He didn't get hurt but he laid there sprawled out on his side with his long legs stretched out. At birth, a horse's legs are about 90% of their adult length. After some time, with some encouragement from those leading the horses, he finally rolled over on his stomach. He got his long legs untangled, finally getting his front legs in front of him and stood up. The parade continued. Horses stand up with their front feet first, whereas cows are the opposite, using their hind legs first. Grampo not only took a lot of pride in his horses but also in the harnesses they wore. As I mentioned earlier, my mom got rather teary-eyed whenever she remembered the scene and told about it, so one can only imagine how grampo must have felt. Although it was a bright, sunny day, for my grampo it was likely the darkest day of his life.

At the end of the day on March 10, 1936, grampo, instead of once being known as the wealthiest man in

Menominee, was now the poorest. At the age of 64 grampo's farming career was over. He and gram moved to a small house in Menominee, IL, that his brother, Louis, owned. He would help farmers with various jobs like assisting in the birthing of calves and many other odd jobs that come up on the farm that require an extra hand or two. His two youngest sons, Nick and Donald, rented the farm from the bank for the next two years following the sale. The farming economy during those two years did not improve at all. In 1938, the farm was sold. Eventually it was sold to a developer and now it is a sub-division. Grampo and gram stayed in their little house in Menominee until they needed care and moved to Galena to live with their daughter, Blanche. Their next move was to St. Joseph's rest home in Freeport. Somehow, my grampo remained upbeat through all of the good and bad years. My gram was the type of person who would smother anyone with her kindness. There is an old saying, "It's not what you gather on this earth, but what you scatter."

FARM FOR SALE

3½ miles from here, 6 miles from there and 2 miles after you get there.

FEBRUARY 30, 1988   STARTS PROMPTLY AT 13 A.M.
TOP HALF WASHED AWAY — CAN BE SEEN AT THE SANDBAR SOUTH OF CARLYLE

ALL FIELDS SUITABLE TO EROSION

Corn makes 25 gallons per acre in a good year
This farm has been operated on a Corn, Corn, Soybean, Ragweed, Popcorn rotation
5 Illinois Bridges makes all fields available (by Pack Mule)

12 Medium to large gullies, doing fine (will be larger by Sale Day)
1 Empty House, door missing, Lean-to, 3 rooms and a path
1 Leaning barn to be moved for lack of storage of manure

GRAIN — HAY

100 bushels of Nubbins (mostly cob)
12 bushels of blight resistant seed corn (should have been planted in 1957)

300 bales of mixed hay (mostly Tickle grass)
100 bales of Ragweed, Sour dock and Cockleburr Hay (no rain)

MACHINERY

1 Poppin' Johnnie (to pooped to pop)
1 Ton Truck with bad transmission (shiftless)

1 Wheel barrow (can trot, pace or gallop)
1 Manure spreader (like new)

LIVESTOCK

1 Running Horse (has tired blood)
1 Man-eating Jackass (sired by night and dammed by everybody)
1 Hound Dog (should have pups by Sale, weather permitting)
1 Holstien Bull (He cow)
1 Jersey cow and 4 calves (this is no bull)
1 Hired Man coming 30 years old

1 Democrat Rooster and 15 Republican Hens
25 Spring Chickens (spring of 1963)
1 21 year-old Hired Girl (some chick)
1 Riding Mare, coming 17 years old (gentle)
9 wormy Pigs and 2 runts
1 Heinz 57 Variety Watchdog (never bothers pedlers)
15 Lambs (10 Marys and 5 not)

MISCELLANEOUS

1 Spiro Agnew Watch (Lots of Tocky but no Tickey)
1 Set of Hens Teeth (Rare)
1 Set Rope Harness (Twine condition)
1 Bath tub with ring included
1 Baby carriage (balloon tires and fluid drive)

1 Arkansas Credit Card (5 gal. can and piece of hose)
50 Hickory and Cottonwood Posts (Seasoned)
And many other items to delapidated and sad to mention

Col. Shoutenslobber - Auctioneer
P.I. Longpencil - Clerk

LUNCH Served by Ptomaine Sisters of Jackass Jct.
Terms 25% Cash, balance 90 days In County Jail

Financing arranged by Last Natl. Bank of Hoot Owl Gap

To Keep Your Farm From Getting Like This

A different kind of farm sale

Chapter 12

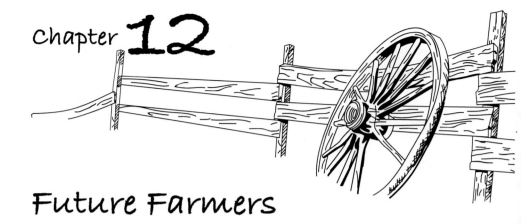

Future Farmers

When Ann and I travel through the Midwest farm country, my places of first choice to stop for breakfast or noon lunch are small town coffee shops or cafes. There, we often hear farmers' conversations and I will look for opportunities to join in. I always wear a DeKalb baseball cap or some farmer-type cap that quickly identifies me as most likely another farmer.

Their conversations cover a lot of territory, from a unique style of homespun humor to the juiciest gossip in town, the price of corn and soybeans, the price of land at the latest farm auction, and the weather is never omitted. An occasional political discussion will occur that will sometimes get a little heated. Added to that list in recent years, is talk about the latest tools of their occupations, robotic milkers, drones, GPS, autonomous tractors, and that list can go on long enough for a second cup of coffee. What they are talking about is no longer science fiction and no longer a dream, if you have enough money and acres.

I once heard a farmer say that changes from the beginning of agriculture to the beginning of the 19th century were so insignificant that if a farmer from that early period came back to earth, he would still know how to farm. A farmer who went to heaven in the 1900s and came back to farm wouldn't have a clue how to operate it. My grampo would fit into that category.

Part 1: Milking Robots

Robotic milking is the automatic milking of dairy cows by a robot to replace human labor and was developed in the early 1990s in the Netherlands due to labor shortages. This system allows cows to be voluntarily milked several times during the day or night, rather than the more traditional time of twice daily. The cows are usually constantly housed in a barn where they have access to feed and water and clean free stalls to lie in. When the cow wants to be milked, she will go into the milking box where she is restrained and the robotic

Robot milker

milker will do its thing while she is fed a ration of feed. The automatic arm positions itself under the cow's utter where it will wash and disinfect it. A laser will then guide the teat-cup to the teats. The milking process now begins. When the utter is emptied, the cow is released and she returns to the well-lit barn where she eats, drinks, sleeps, or roams through the spacious aisles.

In mid-December of 2018, Ann and I visited Willow Valley Dairy Farm near Kent, IL, which is owned and operated by the Lawfer Family. Upon entering the drive of the farm, we met the milk hauler who had just finished loading a day's milking at the farm. I asked him where I might find Ron Lawfer. He wasn't sure but just then, Ron's wife, Julie, came out of the calf barn. We drove ahead to meet Julie, introduced ourselves and said that we were seeking information on robotic milkers for a book I am writing and asked if we could talk to Ron. Julie was most gracious and directed us to the office adjacent to the milk barn where we would meet her son, John, and Ron would be there shortly. We were again very warmly received as John welcomed us in the office. After some explanation as to why we were there, John began explaining the complicated technology of a robotic milking system.

The operation is owned and operated by Ron Lawfer and wife, Julie, and their son John and wife, Elise. Andy Larsen, long-time hired hand (and as John commented, they consider Andy one of the family) also helps to keep the operation going. Ron and Julie's son, Ben serves as dairy nutritionist. Ben, Sam, and Sydney Lawfer are part-time helpers.

In 2014, the Lawfers invested heavily in a state-of-the-art milking facility with two robotic milkers, automatic barn scrapers, a robotic feed pusher, and data tracking system for every cow. The 300- by 124-foot fabric structure was the first of its kind for a dairy. It provides natural light and air flow that John said will stimulate milk production up to 15 lbs. per cow per day. Later, a third milker was added. He also pointed out, as we had noticed, the natural airflow eliminates much of the "cow barn odor" that is usually associated with dairy barns.

As John was showing us the data tracking system on the computer, Ron and Julie entered the office and they were also delighted to answer our numerous questions. It is obvious this is a way of life for them that they love and are proud to be doing together as a family, and enjoy sharing their experiences in this new operation.

Some of our questions:

Q. What is the daily average pounds of milk per cow?
A. We are averaging between 80–85, but we have several cows that will peak their milk production at 150 pounds per day. Along with increased production per cow we feel the cows are less stressed and healthier in our new facility.

Q. What if the milk is contaminated, such as with mastitis or blood?
A. The sensor will detect that. That cow will be milked but that milk will go to a waste pit and the milking unit will be automatically washed and disinfected before another cow is received into the stall.

Q. Are you having any problems or issues with the EPA?
A. We have had an EPA inspection and there were no violations. Before we built our dairy, our plans had to be approved by the Illinois Department of Agriculture. Our facility is designed so there is zero discharge of nutrients into the environment. We do have routine inspections from the Illinois Department of Public Health. Twice a year we are visited by a milk inspector to make sure we meet all the criteria of the Pasteurized Milk Ordinance and are able to ship Grade A milk from our dairy. A federal inspector comes every other year.

Q. What is the greatest advantage of a robotic milking system?
A. A quick answer dealt with labor. The robot is always here on time and doing its job dependably, whereas that wasn't always the case with hired laborers. That means also that we have less problems with mastitis as the robot milks the cows dry. Sometimes the hired labor

would get in a hurry and be careless with their treatment of the cows. Also, they said that the robot never gets angry with the cows when they don't cooperate.

Q. What is the cost per robot unit?
A. $200,000.

Q. Are there cows that won't adjust to robotic milking and do some kick at the robots?
A. Occasionally we have a problem cow. Kicking at the unit sometimes occurs, but the robot is designed so that if a cow kicks at it, they can't hurt it. These problem cows don't stay on the farm long and end up at the packing plant. When doing AI (artificial insemination), we select semen from bulls who are noted for producing heifers that will adapt to robotic milking.

Q. How many times per day will a cow enter the milking stall?
A. Usually three times, however the better producers choose to be milked four to five times per day.

Q. How many cows will one robotic unit service?
A. 50 to 75 cows.

Q. If any part of the milkers malfunction, what happens next?
A. We get a call on our cell phones day or night by a sweet female voice telling us of the problem.

Q. Besides occasionally getting woke up in the middle of the night, are there any other negatives of robotic milking?
A. Our electric bill is slightly higher than when we were milking in our previous facility. However, we are milking more cows. On a per cow basis our electric bill is actually less per cow. The initial cost was a lot of money and milk prices are the lowest they have been in nine years, but if we had to do it over again, I would do it in a heartbeat. The efficiency we have in producing milk, the flexibility in our workload and the overall health of our animals has made robotic milking a wise investment.

Q. We are most intrigued by the tracking system. What are some of its functions?

A. It tracks the milk production and profitability of each cow, clocking her every move from stepping, chewing, eating, to milk production.

Q. Before I asked the next question, I said I didn't want to get personal but do you care to reveal how many cows you milk?
A. No problem, about 200. Right now, we are milking 180 with 20 cows currently dry which is usually the norm. Besides the 200 cows, we have 200 young stock of various ages. Then John proceeded to tell about the crops they raise which consist of 600 acres of corn silage, alfalfa, rye silage, shelled corn, soybeans, and oats, primarily for feed.

Q. What do you do with any bull calves birthed by your cows?
A. We raise our bull calves to about 300 lbs. and then take them to the sale barn for someone else to finish.

My Grampo Kieffer nor my Dad could have ever visualized the size and technology of an operation like the Lawfer's, much less the mega, "Fair Oaks Farms" of Indiana that milk 30,000 cows and the numerous sizes in between.

Part 2: Cattle Ranching

Before getting into modern-day ranching, I want to present a brief history of the beginnings of ranching in the U.S. When Texas became independent from Mexico in 1836, many Mexicans left Texas leaving their longhorn cattle behind. Texas farmers claimed the cattle but there were more cattle than the demand for the meat. The herds continued to grow. When the Civil War

Yesterday's ranching

Today's ranching

broke out, many farmers left their farms to fight the north and the herds continued to expand until 1865, with approximately five million roaming the Texas Plains. After the War, the South's economy tanked along with the demand for cattle. There was strong demand in the North and East. This triggered the notorious cattle drives on the Chisholm Trail from Texas to Abilene, Kansas, the first shipping point to Chicago. Later, other shipping points known as the "End of the Trail" sprang up in other cities in Kansas. In 1866, longhorn cows were worth as little as $2 per head in Texas and $40 per head in Chicago.

As the railroads built southern lines other trails emerged. Numerous challenges arose on the approximately two month drive. Rivers had to be crossed, such as the Red River in Arkansas, and many creeks and storms were a threat. Rivalries with Indians were common, especially going through Oklahoma as at that time, Oklahoma was still Indian Territory. Longhorn cattle were by no means tame and gentle animals which made them prone to stampeding. Two thousand head of cows would be a typical number on a drive, traveling a distance of ten miles on a good day.

The Chisolm Trail was the beginning of Western Folklore. Many movies were made and songs written with many more to follow. The trails ended in the mid 1880s.

On May 20, 1862, the Homestead Act was signed into law by President Abraham Lincoln. This act was intended to encourage western migration by which any adult, male or female, and immigrants were eligible, but citizenship was required. The 1862 Act included black citizens and that gave 160 acres to citizens virtually free.

Later Congress recognized that 160 acres could not support a family in dry land such as the sand hills of Nebraska so they increased the acreage to 320. In 1916 an amendment was added increasing the acres to a full section of 640 acres.

After the Civil War, homesteaders in droves began migrating to virgin soil. This caused friction between the already established rancher and the homesteader. The homesteaders replaced the rangy Texas Longhorns with Shorthorn cattle, a dual-purpose breed for both meat and milk. They are docile with good disease resistance and as they originated in England, they could endure harsh winters. The West was built on Shorthorn Cattle.

Unlike the traditional small livestock farming enterprises of the East and Eastern Midwest, large cattle ranching started to take root in the 1850s and 1860s. Probably the ranch with the most notoriety is the King Ranch of Texas. Started in 1853 by Richard King, it is today the largest in Texas and was notable for developing the Santa Gertrudis Cattle that replaced the Longhorn by crossbreeding, mating Brahman Bulls with Beef Shorthorn Cows. This was the first breed to be developed in the United States. The breed was developed to endure the heat and humidity of the Southwest. The King Ranch was also instrumental in developing the Quarter Horse.

Another large ranch empire known as the Miller/Lux Ranch started in 1863. Henry Miller, a German immigrant and his partner, Charles Lux, had holdings of 1,400,000 acres, plus grazing rights of ten times that much that added up to 22,000 square miles. The ranch land was in California, Nevada, and Oregon where a million cattle and 100,000 sheep grazed. Legend has it that Henry and Charles could ride a horse from Oregon to the Mexican Border and never leave their ranch. After the death of Charles Lux, also a German immigrant, Miller became the sole owner of the empire.

A well-known story of Miller and his top herdsman, Ace Mitchell, was about a cow with Lump Jaw disease. One day Henry and Ace were riding through the vast spread in a cushion-seated, spring wagon when they spotted a Lump Jaw cow standing in the middle of the dusty dirt road. Henry ordered Ace to kill the cow. Ace, for whatever reason, failed to follow orders. About a year and a half later, Miller and Mitchell were riding through the pasture when Miller noticed the same Lump Jaw cow. Mitchell was fired on the spot. Mitchell was later to have said, "30,000 cows in that pasture and that Lump Jaw cow had to be right along the road costing me my job."

Other mega ranches in the U.S. are not necessarily owned by cowboys but investors who rarely sit in a saddle. To name a few, John Malone with 2.2 million acres of land in the U.S. which surpasses Ted Turner's 2.0 million acres. Other than mega ranches there are considerable variations in ranch sizes from a few hundred head to several thousand.

Ranching or raising beef cattle on a smaller scale was going on and still goes on. For some farmers, it is a side-line enterprise with a few cows, especially in areas where portions of their farmland is only suitable for grazing cattle. And then there is the hobby farmer. I know of an investor who recently purchased a recreation farm of 180 acres of mostly timberland, some pasture and a few tillable acres. The last I knew, he had seven cows. Some hobby (or city) farmers have a few horses, usually more than they need, to round up their "big" herds.

Since early ranching, some aspects of ranching stayed the same while others are quite different. The cows are still a machine for harvesting grass. Cowboys round up cows and calves for various reasons. The labor is intense, especially during calving season. Cows in the northern regions have to be fed and prices still have their ups and downs.

The invention of barbed wire in 1874 was the beginning of the end of the open range and the start of many changes in ranching. Breeds of Herford and Angus cattle became the primary grass harvesters. In the early 1970s, other breeds like the Gelbvieh, Limousin, and Simmental to name a few, started to appear. At about this same time, cross breeding began for the advantage of hybrid vigor. Most farmers and ranchers adapt to new agricultural practices. Some are reluctant to change. I recall talking to a rancher whose family raised Herford cattle for about the last 150 years. He wanted no part of cross breeding. He said, with tongue in cheek, I'm sure, "If an Angus bull comes on my ranch and services one of my Herford cows, I'll cross him with a 30-30 bullet."

Cameras can be used to observe cows when calving and notify the rancher by cell phone if the cow needs help in delivering her calf. Drones have unlimited uses: finding lost cows, checking fences, observing water ponds, helping round up cattle, and more. I suppose the horses appreciate drones and modern technology, but then again maybe they are feeling a bit rejected.

There are enormous challenges for ranchers. Like dairymen, they have labor issues, long hours, long days, and living in isolated areas, many miles from a town. Expanding their operations can be quite expensive due to billionaires such as Ted Turner and John Malone with their vast holdings.

Lab grown meat is a growing concern for ranchers and farmers. Will this one day make meat animals obsolete? After the changes of the last 50 years, I can't be surprised by anything. Then what becomes of the vast range lands in the world, especially in the Western Hemisphere—maybe housing for a few billion more people?

Environmentalists are concerned about greenhouse gases that are produced by cows causing global warming. There is a joke going around among cattlemen, "Will we soon have to put diapers on our cows?" The perception seems to be that cows are just one more animal destroying the planet. Will they get the same respect as many of our protected species in the world?

Part 3: Grain Production

The Carroll Family Farms is located about two country miles from Carthage, IL. Nearly all of the farmland is in Hancock County. The family farms also have operations near the city of Luis Eduardo Magalhaes, in the state of Bahia, Brazil.

Late in May, I called Dan Carroll, whom Ann and I had met in a previous farm venture we were involved in. I told him I was writing a book about farming from early days to the present and want to include some large operations happening today and would he be willing to share some thoughts about the Carroll Farms for me to use in my book. He graciously acknowledged he would be willing to oblige. I said we would be passing by his way on the following Saturday on our way to visit our son and to attend our grandson's First Holy Communion taking place on Sunday. Dan said he would not be able to meet us but that his son, John, would be there along with Dan's brother, David, and they would be glad to share what they could.

Driving into the farmstead presented a scene of rural beauty and prosperity. We waited at the new machine shed where Dan advised us to meet. David and John were still at lunch at their respective homes. We were glad to know that this family keeps the tradition of dining with their families when possible.

John soon appeared with a friendly smile and a tight handshake. I introduced our daughter, Tricia, as John had not met her before. We had a brief conversation as he was needed to help with some daily chores. David had finished his lunch and appeared on the scene and after greeting one another, David invited us into the machine shed snack room. We got right into our short interview and a brief history of their operation.

Dan and David's father, Darel Carroll, took over his father's farm operation in the 1950s, gradually increasing the land base. Dan and David entered the operation in the 1970s, accelerating the expansion of the land base, as well as the swine and grain operations as new family members joined. The farm is now in the family for five generations with three generations currently active involving seven family members and more younger members showing interest in getting more involved.

Darel is now 89 years young and still helping with much of the work. He is a unique farmer who lived and farmed through the years of the greatest changes of agriculture. He can tell stories about planting corn with a two-row planter, drawn by horses using check wire that I explained in Chapter 8. He can tell about picking corn by hand and throwing the ears at a bang-board on the wagon. He can also tell stories about operating a 12-row John Deere S680 combine. David told how his dad would occasionally go to Brazil during busy times and help with field work there. David also pointed out how handy it is to have his dad make trips to the implement dealer for parts with his experience as he knows just what they need and how he might be able to improvise if necessary.

After this brief history of the farm, I began to ask questions of their current operation:

Bert: How many tillable acres do you farm here in the U.S.?

David: 8,700 acres mostly in Hancock Co. with a few acres about 30 miles away in Iowa, of which two-thirds is in corn and one-third in soybeans. Our average corn yield is 225 bushels per acre and the average bean yield is 65 bushels per acre.

Bert: Is any of the corn and beans sold or is it all fed on the farm?

David: Our hogs eat every kernel of corn. Our beans are sold but we buy much of the processed product back and maybe more to feed.

Bert: Can you tell us about your machinery, planters, combines, etc.?

David: We have two 36-row planters with 30 in. row spacings. This makes a 90 ft. wide planter. On a good average day, we can plant 700 acres per day. Our two combines are equipped with 12-row corn heads for a total width of 30 feet. Our machinery is primarily John Deere. Our tractors consist of 500–600 HP (horsepower) for tillage operations and 300–400 HP for harvest/planting operations.

Bert: Would you tell us about your hog operation?

David: We have 6,000 sows. It's a farrow-to-finish operation. Previously, we sold some feeder pigs but currently we finish all of them. The litter sizes born average 14½ pigs. The average number weaned is 11½ pigs per litter. We have an employee watching the sows 24/7 around the clock to save as many pigs as possible. The employee doesn't have to save many babies to earn his wages. We have 45 full-time employees here in Carthage. We sell market hogs at 285–290 lbs. which is the preferred weight of the packing plant.

Bert: Do you have any issues with the EPA (Environmental Protection Agency) with this large number of hogs?

David: We are inspected once a year and have always been found to be in compliance. Manure is spread on our crop ground and we keep our pits up to standard.

Bert: About your Brazil farming operations —what motivated you to farm 4,500 miles away from home?

David: To expand here in the U.S. became quite challenging. We wanted to give any family members the opportunity to join if they wished to continue their way of life and to eat a pork chop now and then. In the early 2000s, land was hard to come by and expensive, selling for $3,000 per acre versus $300 per acre in Brazil. The Brazil operation began in 2002 with the purchase of 4,400 acres in Bahia. In 2003, Dan's son, John, after college and marriage moved to the farm as manager. He remained there most of the first eight years. As time went on, expansion increased. We now own and rent 24,000 acres. The majority of the acres are planted to the more profitable crop of cotton. There are approximately 200 U.S. farmers and corporations that have farming operations in Brazil. We also own a cotton gin. John, in addition to managing the farm and 40 employees, also assists other U.S. farmers and corporations in Brazil with their management of finances and regulations in Brazil. Dan occasionally flies to Brazil to fill in or to assist John. When it's winter in the states, it's summer in Brazil which enables family members to keep busy nearly 365 days of the year.

Bert: I asked if it was much of a hassle traveling to their Brazilian farm.

David: From our farm in Carthage to our farm in Brazil that is located 80 miles south of Luis Eduardo Magalhaes, is about a 30 hour journey. Sometimes, it is not fun, it can be frustrating if planes are delayed and such.

Bert: Do you plan to expand operations in Brazil?

David: Probably not, as in 2009, the Brazilian Government put restrictions on the amount of land that foreigners could own and rent which is 3,000 acres. We are grandfathered in but we can't purchase additional land.

Bert: Final question. What do you foresee for the future of farmers and agriculture in general?

David: For a young farmer to start a farming operation on his own is out of the question. Young people will need to have a family farm to take over or to work into ownership with a farmer who has no heirs to take over. Established farmers will continue to have good and bad years, but over the long hall, agriculture will prosper. Agriculture seems to be able to make changes and meet the challenges that come next.

Part 4: A Midwest Mega Grain Farmer

When I asked this mega farmer if I could interview him and use it in my book that I plan to publish, he was very happy to do the interview but did make it clear that he wished to remain anonymous so I accept his wishes and refer to him as Ben. He farms somewhere around 30,000 acres.

Bert: How did your operation begin?

Ben: My grandfather started farming and Dad joined to expand the operation. My dad is still very much involved in our family operation.

Bert: Are you a corporation?
Ben: We are family owned but incorporated.
Bert: How many miles are you spread out?
Ben: We farm within a 100-mile radius in 16 counties in two states. We formerly farmed land in the South where land and rent were much cheaper, such as in states of Texas, Mississippi, Arkansas, Oklahoma, and Louisiana. Besides growing corn and soybeans in the South, we grew milo, cotton, and rice.
Bert: Why did you discontinue farming in the South?
Ben: In 2013, corn was selling for $7 per bushel and soybeans at $16–$18 per bushel, rent skyrocketed and about a year later corn tanked to $3 per bushel and soybeans, likewise, tanked. Also, my kids were getting older and wanted to stay in the North.
Bert: What kind of hassle is it to get machinery, tractors, etc. to the right farm at the right time?
Ben: We transport most of it on semis but occasionally, drive some on side roads. I once drove a tractor and planter to Texas. It is sometimes just as quick to drive as to partly disassemble a planter and load it on a semi. To disassemble a planter for transport is a day's project and a day to put it back together, plus a day on the road, plus more labor cost. It took 2½ days to drive the tractor and planter.
Bert: How do you negotiate crossing rivers with narrow bridges?

Ben: To cross the Mississippi, we use interstates as it is safer. Once across the Mississippi, we use the rural roads.

Bert: Is your current location the largest in this area?

Ben: We are among the largest, but there are some larger, according to what I have been told. Whether we are the largest or not is of no importance to me.

Bert: Of the acres you farm, what percentage is owned or rented?

Ben: About half owned and half rented. Land is hard to buy, but there is always some available to rent.

Bert: How many workers are needed to keep the operation going?

Ben: We have 20 full time employees, two full time mechanics, and we hire a full time agronomist. Help can sometimes be a problem. During wet weather, we have to keep them on the payroll whether we have work for them or not or they will seek other employment. We also hire seasonal help.

Bert: Do you irrigate?

Ben: We have 24 irrigators available and ready to go when needed.

Bert: What is the yield for your corn and soybeans?

Ben: Last year, our corn made 260 bushel per acre. That was a disappointment as we didn't get the fungicide on in time to cool it and it was simply too hot. Our beans averaged 74 bushel per acre. On 300 acres, we had a high yield plot of 114 bushel per acre. With a tight checkbook, we concentrate more on profit per acre rather than the highest yield possible. For example, we find it more profitable to shoot for 200 bushel of corn rather than 250. We start planting in March with beans first so our plants can capture the maximum sunlight. Harvest begins about Sept. 5th and we try to finish by Thanksgiving or early December. We like to start corn harvest at 22–25% moisture. At corn harvest, a truck is loaded every ten minutes. We can dry 4,000 bushel per hour. Our storage capacity is 1.5 million bushels.

Bert: Would you tell me about your machinery?

Ben: We have six 36-row planters with spacings of 20 inches. We have ten combines equipped with 18-row corn heads. Our ten tractors are all four-wheel drives. One of our sprayers holds 5,000 gallons of water that can spray 1,400 acres per day. We also have our own airplane for spraying insecticides and fungicides. I am guessing that we have 30 semi-tractors and trailers.

Bert: Do you have any issues with the EPA?

Ben: One of our biggest concerns is emissions control devices on our tractors and combines. Keeping them up to standards is one of the greatest expenses of keeping them running. That cost exceeds $8,000 per year per unit, plus down time. We have to notify neighbors prior to spraying which is a real hassle since we are so spread out. That means seeing a lot of neighbors and many times, we can't find them at home.

Bert: Do you buy your seed through a sales agent or do you have your own dealership?

Ben: We buy through sales agents. We plant Asgro soybeans and DeKalb corn.

Bert: What are the greatest challenges now and in the future for grain farmers?

Ben: The cost of the inputs versus what we receive for grain. We have to pay a set price to the implement dealer for a bag of seed, herbicides, insecticides, or whatever we need and we have to sell our grain for whatever the elevator is paying. Prices we receive have not kept pace with the input cost.

While doing this interview, Ben gave me a tour of a portion of his operation, his fertilizer plant which has a 7-million-gallon capacity. He said he does custom fertilizing for neighboring farmers. He also showed me the headquarters where most of his machinery is kept. The machinery I saw made the area look like a John Deere implement dealership. The last he showed me was the rock quarry he operates along with his farming operation.

I thanked him for his time and the information and we parted ways.

Part 5: Hog Production in 2019

Recently, a friend of mine, Denny Hermes, one of nine boys raised on a farm near Dixon, IL, told me a story that he remembers hearing his father telling about a common sight in their area when his father was growing up. He said you might see a farmer driving a couple boars on a township or county road to his farm. He would be borrowing the boars from the owner, a neighboring farmer. Borrowing boars was a common practice in those days just as

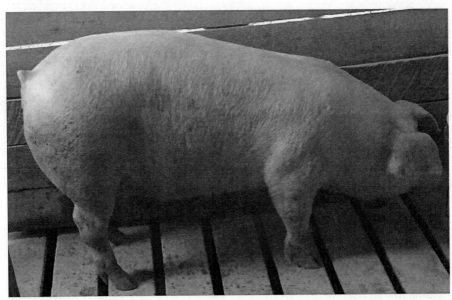

Today's pig

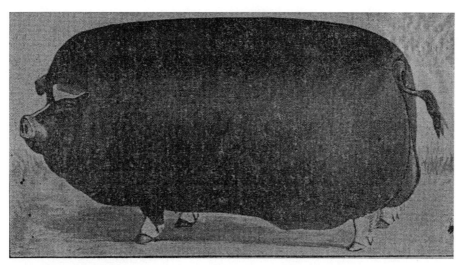

Yesterday's pig

borrowing tools or machinery without charge as it was just the neighborly thing to help one another out when you were able.

We most likely will not be seeing farmers driving boars along a country road any longer since the hog production has moved from small operations for the most part to mega or factory farms. In the last three to four decades, artificial insemination has become the normal practice for impregnating sows. In these mega operations, hogs are raised in confinement numbering in thousands of sows. 100,000 sows or more for one operator around the world is quite common including eleven such operations in the U.S. The largest hog operation is in China known as the W G Group which is the parent company of Smithfield Foods, and has nearly 1,200,000 sows around the world. China produces about half of the world's pork and in order to feed their population of over a billion people and save space, they began housing them in multi-story buildings or "hog hotels" as they call them. Each story can accommodate 1,000 hogs in a building as high as 13 stories.

Raising hogs and poultry in confinement with a few exceptions has become the norm, a practice that animal-rights people frown upon. I am not taking sides on the practice, but merely stating facts about the way modern farming provides food.

Confinement does have advantages. The mortality rate of babies farrowed in crates is much lower than the old way of farrowing in pens or pastures. Farrowing crates help to prevent the mommies from stepping or laying on their babies and even preventing the mommies from eating their babies. "Confinement" to people who don't have real experience with animals, automatically sounds like a "bad" word and it can be. However, confinement of livestock can also provide protection and better conditions for the animal as well as more profit for the farmer, cheaper food for the consumer, and increased food supply for the world.

For finishing operations, hogs live in modern buildings with better temperature control and the automatic feeding and cleaning makes labor much less intense. In addition, they get a dietician diet with a smorgasbord of ingredients mixed with a well-balanced feed. Modern housing, diet, and genetics all play a role in faster weight gain. In grampo's day, the goal was to raise a hog to 220 lbs. in six months. Today's hogs reach 280 lbs. in 5½ months which is the desired weight for packing plants.

Recently, I was at a hog confinement farm visiting with the farmer about pros and cons of confinement. He summed it up by saying, "A hog never had it so good. They don't have to endure extreme temperature changes, the food is good, and their environment is cleaner."

Competition among farmers always existed like people in all walks of life. One example in grampo's time was getting their hogs to market in less than six months. I recall a conversation when two farmers were razzing their neighbor. Uncle Henry, whose hogs needed nearly seven months to reach the desired weight, responded with, "What's the big deal; time means nothing to a hog."

Feral Hogs

Hogs were brought to America with the early settlers. They were an important part of the colonists diet, each person eating at least two hogs per year.

Containing hogs and other livestock on farms and plantations was challenging as wire was not heard of until the late 1800s. Rock, rail, and other such types of fences had to do. Livestock often ran at large. Hogs, when confined, often rooted through and escaped, never to return and are still running wild, mainly in the South and Southwest U.S. to the tune of approximately six million. They became known as feral hogs. Trapping is one of the cheapest and most effective ways of controlling feral hogs. Hunting is widespread but a slow way to eradicate them as they have a high reproductive rate and hardly any natural predators. The U.S. Government offers assistance in eradicating them. They primarily forage for their food by rooting for it, causing farmers billions of dollars in damage. One of their easiest meals is to root along a freshly planted row of corn, hardly ever missing a kernel. The farmer can end up with zero population.

Part 6: Poultry Production

The history of fairs goes back to the time before Christ, although somewhat different from what we recognize today. The fairs took place more often as they served as a marketplace, festival, and all sorts of trade. As today, they also provided social interaction and entertainment.

Before poultry confinement where Ann raised her 4-H chickens

The first Agriculture Fair in North America goes back to 1765, in Nova Scotia. The first fair in the U.S. started in Pittsfield, Maine in 1807. The Elizabeth, IL, fair started in 1820 or 100 years ago. The Elizabeth Fair is a Community Fair, the only Community Fair still in existence in Illinois. All other fairs in Illinois are County Fairs. Jo Daviess County, IL, also has a County Fair in the town of Warren.

In our youth, Ann and I were 4-H members and showed chicken hens and lambs respectively in the annual Elizabeth, IL Community Fair. One of Ann's 4-H projects was the raising of chickens on their farm. At fair time, Ann would carefully scan the chicken house looking for a blue-ribbon bird. From the judges' determination, she never found a winner. She always seemed to take second or third place depending on whether there were two or three contestants. Ann was always mystified as to how the judges picked the best chicken when they all seemed to look identical.

When showing lambs, I fared no better than Ann did with chickens. One year when showing my lamb, I won a red ribbon. Upon returning home, a visitor at our house asked if I won. My proud answer, "I won second place." "How many were in the contest?" was his immediate follow-up question. I was a bit embarrassed when I had to answer, "two."

Going to fairs was and still is a significant event for farm and non-farm youth. Displaying their livestock, poultry, crafts, produce, and such was often the highlight of summer vacation. The competition can be fierce as winning the Grand-Champion steer, hog, or other livestock pays a handsome reward. At the Illinois State Fair in 2019, Governor Pritzker and his wife got into a bidding war for the grand-champion steer, running the price to $75,000. The First Lady of Illinois was the highest bidder. This price, although high, was a bargain compared to the 2016 grand-champion steer that sold for $84,000 to then Governor Bruce Rauner. This is a great contribution to the college fund of the youth showing the champion steer.

Domestication of chickens started about 2000 B.C. From then on, selection for quality continued, especially for layers and meat. They can be found in numerous colors and recently producers of colored eggs have become more common.

Poultry confinement got underway about the same time as hog confinement, bringing along automatic egg gathering. When I was a little guy, gathering eggs was my most dreaded job as I had to sometimes reach under a hen sitting on a nest which would result in my getting a mean peck on the hand. I would also have to keep an eye peeled on roosters and their spurs. Roosters use their sharp spurs to vigorously defend their flock against predators. I was afraid they would take me as a bad guy.

In recent years, some cities changed their animal ordinances to allow backyard chickens but no roosters were allowed. Neighbors didn't appreciate the early dawn wake-up calls. The idea of having their own "free" eggs got some city dweller enthused about raising their own flock. A good hen can lay 276 eggs per year according to USDA figures. A good idea but realistically not so cost effective. After building a chicken house and supplying it with feeders, waterers, and purchasing the baby chicks, feed and medicines, the eggs don't look so "free." Then there is the cleaning and maintenance of the coup and arranging for care of the chickens while away from home. The idea begins to seem a little less exciting.

In spite of large confinement operations, many farmers and ranchers continue to maintain their own flocks, primarily for their own use and sell their extra eggs to their neighbors. They have an advantage over their city cousins as they have an existing building and cheap feed. Some ranchers living in isolated areas, besides keeping a few chickens, will keep a milk cow. Another advantage is a farmer or rancher can keep roosters which are necessary to fertilize eggs so they can hatch out their own baby chicks. Up to 15 eggs can be placed under a clucky hen and in 20 days, little chicks will begin pecking their way out of the egg shells. This is usually about a good day's (24 hour) project. Two hours after exiting the egg, they dry off and become one of God's cutest, soft, little, fuzz-ball creatures.

Some interesting facts about chickens:

Chickens are either classified as layers or broilers.

Their meat is high in protein.

They are the best converters of feed with a ratio of 2 lbs. feed generates 1 lb. meat. Fish are nearly equal to chickens. Cattle in feedlots are nearly the worst with a ratio of 8 lbs. of feed for 1 lb. of meat while cattle on grass are even worse.

They are easy prey, especially for foxes who find them to make a good meal.

They are not so good at crossing roads. However, of all the domesticated fowl, ducks are the worst as it would be almost suicide for them to wobble across a road, unless a road block was in place.

Roosters have a unique feature with their crowing. They serve as an inexpensive alarm clock for folks, like farmers, who want to rise at the crack of dawn. Hens sound out with loud clucks after laying an egg to announce news of food production.

Chickens, like most fowl, do not have teeth, thus the expression, "As scarce as hen's teeth."

They do not have stomachs so they have a gizzard that serves to grind up their food. They need grit or small pebbles to help in the grinding process. The chicken's bill is a great asset at picking up pebbles when grit is not available, as well as scavenging for food. I find gizzards to be very tasty, but some people find them hard to swallow, literally. If you have never eaten one, I highly recommend you try one.

Financially, chickens have a fast turn over as layers start laying at about five months.

Chickens sleep on roosts. A roost is a series of small poles or 2 inches by 2 inches boards nailed across two slanted boards, 2 inches by 6 inches that are attached to the high side of the chicken house and extending out at the bottom so that one row of boards is not directly over the row below it. Chickens like to sleep on roosts as a natural instinct to avoid sleeping on the floor to avoid lice and mice as well as predators.

Broilers were noted for providing the entrée for major farm activities such as threshing, filling silo, shredding, etc.

Chickens are noted for many different breeds, more than any other species topping sheep that number nearly 1,000 different breeds worldwide.

Henry A. Wallace, 1888–1965, a native of Ames, IA, was recognized as a Farmer, Secretary of Agriculture, U.S. Vice President, Secretary of Commerce, co-editor of *Wallace Farmer Magazine,* and Co-founder of Hi-Bred Corn Company which in 1935 became known as Pioneer Hi-Bred corn Company.

In addition to the above list of accomplishments, Henry A. Wallace with his son Henry B. Wallace, recognized the impact of hi-bred corn and began research on Hyline chickens which proved highly successful. These chickens became the best layers with minimal cost and have become the most popular in the U.S. and around the world.

A tale worthy of note about Henry A. Wallace is that while a Democratic politician, he became unpopular in his own party and with many Southerners because of his liberal views. Once while speaking in a southern state, he was greeted with angry protesters throwing tomatoes and eggs at him—eggs of his own developing—the Hyline. Apparently he was comfortable being around his eggs given that a picture shows him continuing to give a speech even with a partial egg shell still lodged in his hair.

Part 7: A Word About Sheep

Biblical history of sheep goes back to Adam and Eve. Sheep are mentioned in the Bible 500 times, making them the most written about animal in the Bible. According to Biblical history, Abel, the second son of Adam and Eve, was the first shepherd. Abel is known as the first murder victim of the human race.

A flock of sheep grazing in NW Illinois

Domestication of sheep began about the same time as other plants and animals—10,000 years ago. Sheep were no doubt one of the easiest creatures to domesticate due to their docile nature. Actually, because of their docile nature, they have often been thought to have very little brain power, and my experiences with them have led me to believe that they are definitely short on brain power.

As humans spread to all parts of the earth, sheep did likewise, making their way to America on the second voyage of Columbus in 1493. In the U.S., sheep production hit a high of 51 million head in 1884, but are down to only about six million today. The decline is caused by the decline of family farms with more diversified livestock operations and due to the drop in the demand for wool with the development of synthetic fabrics. Raising of sheep came to be replaced by the more profitable raising of cattle. Sheep are still a thriving enterprise in the countries of China, Australia, New Zealand, Sudan, and India.

As sheep and cattle ranching developed in the Western U.S., much friction developed between the sheep ranchers and cattle ranchers. Competition for land and grazing rights became fierce, resulting in range wars. I am not going to elaborate on the range wars as that can be a story and a half for another day.

When I was a 4-H member, sheep were my project. I had to learn a lot about them such as how to properly care for them, sheer them, and become familiar with the popular different breeds in my area. They consisted of Cheviot, Dorset, Hampshire, Suffolk, and Whiteface. These are just a few of the 60 different breeds in the U.S. and an estimated 1,000 worldwide.

Sheep are noted to cling together, staying in one flock. In spite of the infamous Bible parable of the "one lost sheep," for one to stray from the flock is a rare occasion, unless they have been frightened by a predator. Otherwise they tend to follow their leader. When I raised sheep I could lead them anywhere with a bucket of grain. All I had to do was get one to follow and the entire flock would tag along. Sheep have one, two, and sometimes three lambs yearly after a five-month gestation period. Triplets can be a challenge for sheep producers as the ewes have only two nipples. One of the lambs has to be adopted out to a ewe with just one lamb. This is sometimes challenging as often the ewes won't accept a strange lamb and with a good sense of smell are quickly able to know their own lamb. The other choice is to bottle feed the extra lamb. Bottle feeding lambs was an added chore during a busy season coinciding with the calving season. I would usually delegate the bottle feeding to my kids. Sometimes a ewe doesn't give sufficient milk for her lambs making the need for supplementing with bottle feeding.

In the last two or three decades, coyotes have invaded Northwest Illinois, and are the sheep's worst enemy. Sheep are vulnerable to predators such as foxes, wolves, coyotes, feral hogs, and wild dogs. Sheep have minimal defense as about all they can do is run and not usually fast enough to outrun their attacker. Wolves once roamed nearly all parts of the lower 48 states until around the early 1900s, when they were nearly exterminated. In 1973, they were put on the endangered species list by the Fish and Wildlife Service. Beginning in 1995, much to the regret of cattle and sheep ranchers, wolves were brought from Canada into Yellowstone National Park. This presents a problem for ranchers as wolves obviously do not stay in the park.

When my grampo was a farmer, the howls of the native North American coyote were never heard in Northwest Illinois. In the 1970s and 1980s, the coyote appeared by crossing the Mississippi from the west. Their original domain was the Great Plains and farther points west.

Coyotes are closely related to wolves and equally despised by farmers and ranchers. In Northwest Illinois, farmers who raise sheep need to keep their sheep in a coyote-proof barn at night.

Chapter 13

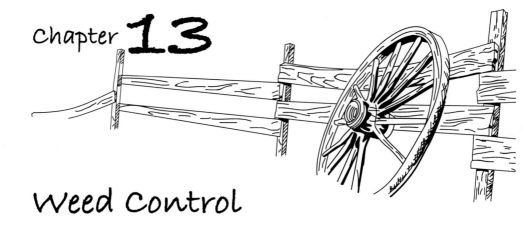

Weed Control

Issues with weeds have been in existence since Adam and Eve as seen in the Book of Genesis, Ch. 3, V.17 & 18 (St. Joseph New American Bible)

> ... *"Cursed be the ground because of you! In toil shall you eat its yield All the days of your life. Thorns and thistles shall it bring forth to you, As you eat of the plants of the field."*

Besides this, there are 29 other verses that mention weeds. Man fighting weeds has been going on forever.

It is important to note that one definition of weeds is "unwanted plants." Some plants (weeds) are a "thorn in the side" to farmers and gardeners while

Weed zapper

these same plants may have medicinal value or be considered a delicacy food item. Dandelions and Lamb's Quarter are two such weeds that many people harvest for eating while the gardener and farmer work hard to destroy them. I remember my mom telling about dandelions being my dad's favorite vegetable. I also recall, a few years ago, when an agronomist was checking my corn fields and when he would come upon Lamb's Quarter growing, he would strip off leaves and eat them.

For the last 10,000 years, there have been primarily two kinds of weed, or unwanted-plant, control. Until the herbicide age that started in the 1940s, control of weeds was hard, manual labor for both man and beast. Before herbicides, methods of ridding weeds had few changes, thus, there is little to write about from the beginning of farming until the 1940s in regard to weed control. Weeds have been a nuisance ever since man domesticated plants. Just hoe, hoe, pull, pull, cut, cut, or burn. Even though the methods changed little, the tools used to control weeds have steadily improved. Primitive man's first tools were, no doubt, a crude hoe or knife. With the beginning of animal power, such as oxen, horses, or camels, high technology in weed control began to advance. Soon more sophisticated tools such as harrows, plows, and cultivators appeared. The first harrow was nothing more than a large stick with pegs driven through it and a handle attached.

In ancient times, there were other methods of killing weeds. One was the use of salt but it also killed crops. The Roman soldiers were noted to put salt on the crops of enemies in an effort to starve them. Amurca, a by-product of olive oil, was also used as a weed killer. It was effective in weed control around the bases of olive trees and other trees.

Implements for controlling weeds are still in use today but to a much lesser extent as herbicides began to replace them. 2-4-D was the first to debut in the early 1940s. Atrazine was registered for use in 1959, becoming one of the farmer's choice herbicide asset to control pre- and post- emerged broadleaf weeds and grasses in corn and sugarcane. In 1973, Monsanto began marketing Roundup herbicide, which contains glyphosate as the active ingredient. As was the case with many scientific discoveries, serendipity was involved with the discovery of glyphosate. When glyphosate was first screened for its ability to control weeds, no activity was observed initially, since it is a relatively slow-acting herbicide. The plants that had been treated with glyphosate should have been discarded, and the molecule would have gone undiscovered. Apparently, however, the scientist that was supposed to throw the plants away went home for the weekend. When he came back on Monday, he noticed the plants had started to die. Because of this chance event, glyphosate went on to become the world's most valuable herbicide. Roundup is also an unusual herbicide in that it kills almost all plants. Most herbicides affect only certain plants. Atrazine, for example, can be sprayed in a cornfield to kill weeds, and it won't hurt the corn.

In 1996, soybeans became genetically modified to resist Roundup, and corn in 1998. By 2005, 87% of soybeans planted were glyphosate resistant varieties. Recently, Roundup Herbicide has been getting some bad press as green extremists oppose glyphosate (the active ingredient in Roundup) claiming it causes cancer, however scientists at the World Health Organizations, the Food and Drug Administration and the Environmental Protection Agency confirm it is safe for consumers and farmers.

About this time, in the early 1980s, biotechnology was emerging as a new branch of science. Biotechnology provided scientists with a way to alter the genetic makeup of plants. Monsanto scientists used biotechnology to put a gene from a microbe into plants that would make the plants immune to glyphosate. The first such plants, called Roundup Ready, were commercialized in 1996. Farmers loved how Roundup Ready crops made it easy to control weeds. Before Roundup Ready crops, farmers had to pick and choose the right combinations of herbicides, depending on which crop they were growing and which weeds they had in their fields. With these new crops, farmers could simply spray Roundup to kill all the weeds in their field and not worry about the herbicide injuring the crop. Within just a few years after commercialization of Roundup Ready soybean, they comprised the majority of all soybeans planted in the U.S. In fact, there is probably no other technology in the history of agriculture that was adopted faster than Roundup Ready crops.

Given the great success of Roundup and Roundup Ready crops, you might assume that weed control is no longer an issue for farmers. Unfortunately, however, Mother Nature plays by her own rules. In the same way that various microbes that cause human diseases are increasingly evolving resistance to antibiotics, weeds have evolved resistance to Roundup, as well as to numerous other herbicides, so the fight between man and weeds goes on. Dicamba is the herbicide that currently is being relied upon to control weeds, but how long before it loses its punch? Scientists are increasingly looking at non-chemical strategies to control weeds. In organic cropping systems, weed control is mainly achieved by using tillage, rotating crops to keep different types of weeds at bay, planting cover crops, and manual weeding. This is fine for smaller acreages, but these intense management practices that are required would be difficult to scale up for the multi-hundred-acre farmers. Perhaps someday small robots will travel up and down rows of corn, looking for weeds and destroying them. Or maybe drones will fly over the crops, find the weeds and spray just the weeds with a mix of potent herbicides. Biotechnology could also be used to develop novel weed control strategies.

Waterhemp is one of the worst weeds in the cornbelt, and has evolved resistance to Roundup and several other herbicides. One of the reasons waterhemp is so good at evolving herbicide resistance is because it is a dioecious plant. That means some plants are males, others are females, so they have to cross with each other to produce seed. Because waterhemp plants cross with each other every year, they exchange genes, which means they can evolve resistance to herbicides quickly. What if scientists figured out a way to use waterhemp's dioecious nature against itself? Could one use biotechnology to make waterhemp produce pollen that would lead only to males being produced? If so, with each passing generation, there would be fewer and fewer females, so fewer and fewer seeds would be produced. These forms of weed control might seem far-fetched, and maybe they are. But necessity is the mother of invention. If herbicides keep failing due to resistance, something will have to be invented.

In Australia, a farmer named Ray Harrington, who was having difficulty controlling ryegrass in his wheat fields, invented what has become known as the Harrington Seed Destructor. When wheat is harvested, seeds of ryegrass plants are pulled into the combine as well, separated from the wheat, and dispersed out the back of the combine. Mr. Harrington devised a machine that was pulled behind the combine, and the chaff, including any harvested ryegrass seed, are sent through the machine. The machine uses a cage mill to grind up the seed to kill it. Consequently, fewer seeds are dispersed onto the field, reducing the infestation of weeds for next year's crop. If farmers relied solely on this so-called harvest weed seed control, weeds would evolve resistance to it. For example, a weed could evade this tactic by having its seeds fall off the plant prior to harvest. This is an example of evolution in action, and highlights that controlling weeds, or any other pests, is akin to an evolutionary arms race. The farmer uses a tactic, the weeds evolve against that tactic, and the farmer uses a new tactic.

One of those new tactics is an electric weed killer, which is gaining momentum as a result of recent pesticide litigation and weed resistance to herbicides. It is also expected to be a less expensive way to control weeds. The electric weed killer super heats weeds by sending an electric current down the stem and into the roots. The machine is hooked to the back of a tractor generating electricity by the tractor's PTO shaft, producing electricity across the width of an electric bar, held just above the crop canopy. It also has discs evenly spaced that go in the ground. Electricity travels down any stems that contact the bar, and completes the circuit by going through the roots and back to the discs. This is a most effective way of killing rhizomes. The electric weeder is a great asset especially to organic farmers. The war between man and weeds will probably go on forever.

Certain weeds are known as noxious weeds, meaning that they are listed by the USDA or a government authority as being a harmful weed. Of the numerous noxious weeds, some worthy of noting are Kudzu, Canadian Thistles, and Poison Hemlock.

Kudzu, an invasive plant primarily thriving in the southern U.S., has become known as the vine that ate the South. It was imported from Asia in the late 1800s as an ornamental plant and its ability to shade porches and entire houses. The vines have modes, forming stems that attach to nearly any surface. Because of its ability to spread rapidly, the vines can spread up to 60 ft. per year, conservationists advocated its use as a weapon for erosion control during the dust bowl. This may have been a good plan, but it soon backfired. As the vines grew rapidly, up to one ft. per day, it soon got out of control and smothered other plants, trees, and everything in its path. It is an edible plant for livestock which helps to control it, but the best weapon against it is glyphosate.

Canadian Thistles are common in Canada and the U.S. Why they are called "Canadian" Thistles remains a mystery as they are native to Europe. The seed, unintentionally, came to America mixed in with other seed imports as early as the 16th century. In pasture and range land, this weed is farmers' and ranchers' most hated weed, mainly because of the difficulty of exterminating it. With the introduction of herbicides, control became easier but still challenging because of its rhizomes. Rhizomes are roots that grow from the mother plant extending horizontally underground. These root systems have nodes that produce other roots several inches apart. Therefore, if the mother plant is killed, their underground nodes still produce a new plant, thus killing a Canadian Thistle with herbicide usually takes more than one year. Plants that shoot stems, like rhizomes, on top of the ground are called stolens. Strawberry plants are a good example of stolens.

Poison Hemlock is one of the most toxic plants to people and livestock. It originated in Europe and North Africa and spread to the U.S. and numerous other countries. It was introduced to the U.S. as an ornamental (I really don't find it that pretty) having white flowers on light purple, hollow stems. Eating just a few leaves can be deadly. The small hollow stems have been known to be used by kids as pea-shooters resulting in possible death.

This noxious weed which flourishes along streams and wet lands, first appeared in Northwest Illinois within the last decade. Although deadly to cattle, they will not eat it unless there is no other forage. There has been an incident in Jo Daviess Co. in Northwest Illinois where a flooded river surrounded an island, stranding cattle with nothing to eat but Poison Hemlock. They all perished. As a bi-annual plant, it is easy to control with herbicides such as 2-4-D.

In recorded history, Poison Hemlock leaves were used to brew a poisonous tea. The famous Greek Philosopher Socrates, 470 to 399 AD, met his fate drinking such tea in the year 399 BC. This was a common way of rendering the death sentence in that era. Forcing a criminal to drink this tea was a common way of Greeks rendering the death sentence. They were generally opposed to crucifixion which was the way of the Romans.

Epilogue

Technology causes many more changes in farm country than just the way we provide food. Larger operations continue to grow leaving fewer small family farmers and ranchers in our rural communities to patronize the hardware store, the veterinarian, feed stores, and such businesses. The decline in the number of farms along with the decline of family size results in fewer church goers as well as fewer school kids, fewer people to serve on schoolboards and various farm and community committees and the list goes on.

Some facts about family farms follow with statistics from the USDA's Economics Research Service. Of the few more that 2 million farms in the U.S.,

Wood-burning Kitchen Stove

the annual gross income ranges from 1 thousand to 5 million dollars. Family farms of which some are incorporated, make up to 98% of all farms and provide 88% of production. Small family farms produce 21% of production while midsize and large farms (farms with more than $350,000 in annual gross revenue) produce 66% of production. Non-family farms are 2.1% and produce 12% of production.

Of the small to midsize farms, 80% of the principal operators have off farm jobs, as well as, 62% of their spouses, primarily to provide for health care benefits.

Adding to rural woes besides urban development are hobby or recreation farms bought by wealthy, wanna-be farmers who never get dirt under their fingernails and are more interested in collecting farm subsidies and a place for finding good tax deductions than to provide healthy food for our tables. Even though farmers are the primary buyers of land, the city dweller or wealthy investor doesn't make it any easier for a young man or lady who truly desires to farm the land to start a farm operation or to expand their operation as the price of land gets driven up by the wealthy to a point where a farmer can't make owning the land profitable.

Technology has eliminated numerous chores that kids were responsible for doing that contributed to teaching them good work ethics and the importance of making contribution to the family unit. I think I have the experience to talk about these chores as much as anyone and I certainly did my share of them. A sampling of those chores is carrying in the wood to fill two wood boxes, before and after school during the winter months, carrying out the ashes, carrying in water from the cistern to fill the kitchen-stove reservoir and a fresh bucket of well-water for drinking. On extremely cold nights there would be a thin layer of ice on top of the bucket that had to be broken up by the dipper that everyone used to drink from. A cold drink was always available in the winter.

On our farm, as I was growing up, we had to walk the approximate 20 yards from the house to the pump, pump the water from the well to fill the bucket and carry it back to the house. Some farms had windmills or gasoline engines but not on our farm. I guess this was because we had a creek, or "crick" as we always called it, nearby where most of the livestock would go for a drink.

Usually the girls' job was to gather the eggs and candle them which was usually a Saturday job. The waterers in the chicken house had to be dumped out at night to avoid freezing, then filled again in the morning before leaving for school. Other poultry such as geese and ducks had to be driven to their sheds at night and the door secured to keep predators out. It wasn't necessary to drive the chickens in as they hit the roost early on their own.

The sows had to be let out of their pens for water and feed morning and night. While they were out, the pens would be cleaned.

Shoving silage out of the silo was always a kid's job, if the kid was brave and old enough to climb the high silo. Usually two of us would do this job, each using a six-tine pitch fork. We would usually shove off a depth of three to four inches at night which would be enough for the morning and night feedings. The silage consisted of corn kernels and cobs in lengths of ¼ to ½ inches long which would collect on the fork tines and have to be cleaned off periodically.

The above mentioned is a partial list of kid's jobs in my youth during the winter months after school. If we hurried up with chores in the evening, we might still have some time for games such as monopoly, checkers, chess, or cards. During the summer months, some chores were not necessary, but were quickly replaced by other chores connected with gardening and raising crops. We worked hard but we seemed to have fun and enjoyed our lifestyle for the most part. Growing up on a farm provided a life-time of memories. We lived a simpler life that was closer to nature. Technology has made the work load easier and I am grateful for that.

I realize that today's specialized farming operations, especially dairy farms, have chores for kids such as bottle-feeding calves, moving them from one pen to another and cleaning out the pens. Some nine and ten year old kids operate bobcats while moving hay around. The kids learn how to operate tractors and pickups at a young age and can do that at a young age if they are responsible and taught to do so safely. I recall our youngest son driving my pickup around the farm when he was barely tall enough to see over the dash. Our oldest was raking hay with a tractor at a young age as well.

Today, the farm youth are kept busy on off-farm activities much more than when I was growing up. They're involved in sports of all kinds. They have 4-H and FFA projects and these are all good character-building activities.

4-H is a youth organization started in 1902 in Ohio. The goals of 4-H are to develop citizenship, leadership, responsibility, and life skills. The 4-H symbol stands for Head, Heart, Hands, and Health and their model is to "Make the best better" with their slogan being, "Learn by doing." 4-H includes projects for boys and girls working with animals and plants and also includes sewing, crafts, mechanics, and so many more.

FFA, Future Farmers of America, also a youth organization, began in 1925. Its original purpose was to teach boys to be better farmers as well as to be leaders in the agricultural world. It started out with only accepting boys but girls began to be included in 1969. The FFA motto is "Learning to do, Doing to Learn, Earning to Live."

These programs began as it became obvious for the need of education in many other areas than farming the land. They attempt to foster an awareness in our youth of the importance of faith in God, stewardship, leadership, citizenship and the need to become involved in our communities and our government to keep agriculture as a vital and respected part of our world. In grampo's time, or in my aunts and uncles day, farming didn't allow for a great deal of time spent on off-farm activities for the young or the old and it was probably not so necessary to have time for that. However, with the technology today, farmers do have more time to be involved and there is a greater need to have a voice and to have some influence in the world around us.

I often ask myself if I would be happier and more fulfilled had I grown up in a modern world. I wonder if technology does enhance the family structure which is still the world's most valuable asset.

Is our technology and our modern way of living responsible for the decline in family life? Certainly, there are pros and cons to these questions. I believe that technology is here to stay and grow even more, so it is going to require us to realize we can become better for it or we can abuse much of it and become much the worse. Some of Ann's and my kids have jobs not necessarily unrelated to agriculture, but have moved four to eight hours apart by car, so we keep connected with Snapchat. We find it to be a wonderful tool. But then I recall some friends of ours who wanted to spend a summer evening at our farm enjoying God's nature in the beauty of the late sunset listening to the many different sounds and it was good until they pulled out their phones and iPads and became totally distracted by their technology. Technology is here, the good and the bad of it and it won't go away, so it is up to us to use it without it overcoming us.

It is evident that farmers will have to produce a massive amount of food to feed an increased number of people. Some estimate over 9 billion in 2050, up from over 7 billion plus today. Not only the increase in population but the increase in the world's wealth and healthier diet will demand that farmers produce more food on fewer acres. Urban sprawl, more roads, hobby farms, recreation areas, and much more will continue to gobble up productive farm acres. At farmland auctions, I've heard the auctioneer say many times as a sales pitch, "Buy it now as they are not making any more." As more food is needed to feed more people on fewer acres, one might conclude that those who produce food would become and remain some of the wealthier folks on God's earth. If history repeats itself, and it usually does, agricultural cycles will repeat themselves. Some recent up cycles were the Roaring '20s, the 1950s following WWII, the 1970s when a farmer couldn't do anything wrong to haul in a good profit, and more recently 2008–2010 when a semi-load of corn fetched as high as $7,000. During that time, it was reported that a large grain farmer donated a load of corn to charity. It is a fact that farmers are also extremely generous in sharing their good fortunes when they have them.

All of these up cycles were followed by severe down trends, sometimes resulting in serious consequences. The farm crisis of the 1980s was one of the hardest in recent memory, even though in many cases that crisis could have been kept to a minimum. During the early years of the 1970s world stocks of

Epilogue 111

grain shrunk, especially in Russia and stock piles in farmers' grain bins in the U.S. were nearly empty. I recall how then Secretary of Agriculture Earl Butz, was encouraging farmers to plant from fence row to fence row. As you can imagine grain prices skyrocketed and farmers followed Butz's advice. Butz also advocated for farmers to get big or get out, thinking that larger farms are more efficient. As the farm economy improved through the 1970s, bankers were very generous at making loans, too generous, as farmers expanded buying more land and larger machinery. The human nature of greed took hold as it always does with high prices. Farmers were thinking this boom would last indefinitely, and they had encouragement in that thinking by many advisors. There was soon a rude awakening for those who went too deeply in debt. As the farm crisis of the 1980s set in, grain farmers became victims of their own success with over-producing, following Earl Butz's advice of planting fence row to fence row. Sky rocketing prices of the 1970s were the biggest culprits as over production soon became burdensome. Beginning in the 1980s, land prices started to tank along with grain and livestock prices. To add to their woes, interest rates soared to near 18%. Many farmers simply couldn't make payments and the additional land purchases went back to the original owners with many of the buyers forfeiting their down payments. The banks that made their too-generous loans to farmers suffered pain as well. As the 1980s farm crisis worsened, many farmers soon found it virtually impossible to make a living off the land much less service their debts. Off farm employment became a necessary option, however jobs were hard to come by as the entire nation was in a recession.

The drought of 1988 was another hardship. A few sprinkles of rain would be cause for a celebration. Speaking of rain, farmers had advantages over some other businesses such as golf course owners, etc. who are also dependent on the rain. During Sunday church services, congregations would offer prayers for farmers to get their needed rain. Of course, you often hear congregations call for prayers for farmers if rain is needed, or they need for the rains to stop for a time, if it's too cold or too hot. One doesn't often hear prayers calling for favors for the golf course owner or the ski lodge owner. The weather seems to be, especially among farmers, every day kitchen table conversation.

Besides prayers, farmers today receive subsidies that most farmers in grampo's day did not. Subsidies started with the Agriculture Adjustment Act that President Franklin Roosevelt signed into law in May of 1933. This became known as the A.A.A. which helped farmers get through the Depression and the Dustbowl of the 1930s. However, farming the land didn't become profitable until WWII started on Sept. 1, 1939, when Hitler invaded Poland.

As German submarines were attacking or trying to attack American supply ships to European Allies, President Roosevelt signed into law on Jan. 30, 1942, The Emergency Price Control Act under which food rations began in the

112 Changes in the Good Life

803 768 EL

UNITED STATES OF AMERICA
OFFICE OF PRICE ADMINISTRATION

WAR RATION BOOK TWO
IDENTIFICATION

Constance Theresa Tranel
(Name of person to whom book is issued)

_____
(Street number or rural route)

803/768 EL

_____ _____ ___ __
(City or post office)   (State)   (Age) (Sex)

Issued by Local Board No. 6243-1 Jo-Daviess Illinois
                                    (County)   (State)

_____ Galena
(Street address of local board)   (City)

By Virginia M. Himmel
(Signature of issuing officer)

Signature ___Constance Theresa Tranel_____
(To be signed by the person to whom this book is issued. If such person is unable to sign because of age or incapacity, another may sign in his behalf)

WARNING

1. This book is the property of the United States Government. It is unlawful to sell or give it to any other person or to use it or permit anyone else to use it, except to obtain rationed goods for the person to whom it was issued.
2. This book must be returned to the War Price and Rationing Board which issued it, if the person to whom it was issued is inducted into the armed services of the United States, or leaves the country for more than 30 days, or dies. The address of the Board appears above.
3. A person who finds a lost War Ration Book must return it to the War Price and Rationing Board which issued it.
4. Persons who violate Rationing Regulations are subject to $10,000 Fine or Imprisonment, or both.

OPA Form No. R-121   16—30853-1

My sister's war ration book

INSTRUCTIONS

1. This book is valuable. Do not lose it.
2. Each stamp authorizes you to purchase rationed goods in the quantities and at the times designated by the Office of Price Administration. Without the stamps you will be unable to purchase those goods.
3. Detailed instructions concerning the use of the book and the stamps will be issued from time to time. Watch for those instructions so that you will know how to use your book and stamps.
4. Do not tear out stamps except at the time of purchase and in the presence of the storekeeper, his employee, or a person authorized by him to make delivery.
5. Do not throw this book away when all of the stamps have been used, or when the time for their use has expired. You may be required to present this book when you apply for subsequent books.

Rationing is a vital part of your country's war effort. This book is your Government's guarantee of your fair share of goods made scarce by war, to which the stamps contained herein will be assigned as the need arises.

Any attempt to violate the rules is an effort to deny someone his share and will create hardship and discontent.

Such action, like treason, helps the enemy.

Give your whole support to rationing and thereby conserve our vital goods. Be guided by the rule:

"*If you don't need it,* DON'T BUY IT."

☆ U. S. GOVERNMENT PRINTING OFFICE: 1942   16—30853-1

Epilogue

following spring. Under the system all men, women. and children were given a food-ration book. This brought bartering back into play. As an example, my family living on a farm could trade vegetable stamps for flour stamps. We had an ample supply of vegetables from our garden but as a large family, we ate a lot of bread using a lot of flour. Sugar stamps were one of the hottest items and they continued until 1947, after the rationing had ended. Sugar stamps weren't so crucial for my family as we substituted honey for sugar as much as possible. Roy Monhanky kept a large colony of honey bees on our farm. Every year to compensate for the use of the bee yard, he would give us a year's supply of honey.

Victory Gardens became a popular term by which the USDA encouraged families, rural and urban alike, to grow their own produce. Likewise, the USDA Extension Services, which had begun in 1914 as a result of the adoption of the Smith Lever Act, followed up on the Victory Gardens encouragement by promoting home canning. I talked about home canning in Chapter 1.

Farming is one of the oldest occupations of man and woman. To provide food from the time of hunting and gathering to modern day farming, it has always been challenging. Today, farmers face new challenges every day such as government regulations, global fluctuating markets, difficulty in expanding their operations because of high-priced land and its unavailability, the high cost of machinery and repairs, weather issues, flat prices, and the list goes on.

In the minds of many, there is a dim outlook for the future of the small family farm of today. The traditional family farm as my grampo, my dad, and I knew it, i.e., raising multiple commodities, is history for the most part. It no longer makes financial sense to raise a few hogs, milk a few cows, fatten a load of cattle. Raising poultry can be a good hobby but unless you have thousands of birds, the profits are quite dismal. The traditional 160 acre family farms of old which made a living for the family have been purchased by a mega farm operation or swallowed up by urban spread. Even those family farmers who expanded their acreage to an average size of 441 acres today, according to the 1917 USDA Ag Census, have some financial realities with which to deal.

In the last few years, over-production with increased input cost and lower commodity prices translates into non-profit and in some cases losses. One bright spot is land inflation of recent years which gives the farmer some equity for borrowing money, for as long as that trend lasts. It becomes quite clear as to why the dim outlook. The U.S.D.A.'s latest Ag census released in April of 2019 showed that there was a decrease of 160,000 farms from 2007 to 2017. It used to be that farms were passed down to the children for several generations, but today that is becoming more difficult as well. Unless the next generation has enough money to expand the present farm operation, they look to a profession or career where it is possible to make a living. The farm becomes another part of a mega farm or an investor project. Another bright spot for farmers, home buyers, auto buyers, and anyone else needing to borrow money is low interest rates. In the summer of

2019, interest rates were the lowest they have in the last 5,000 years—yes, that is not a typo, 5,000 yrs. The highest interest rates were 20% in 3,000 B.C. in Mesopotamia. In more modern, capitalist times, interest rates skyrocketed to 18% in 1982. This is a figure that a lot of farmers will long remember.

In times of financial stress, farmers often blame their problems on such things as low prices, bad weather, or high interest rates and bankers. Legend has it that at the trials of Frank James, brother of Jesse, the notorious outlaws, the jurors were more sympathetic to the bank robber, Frank, than to the banks he robbed and the murders he committed. An impartial jury was not to be found. At three different trials, he avoided convictions and was set free. They were saying that if Frank James wants to rob banks where excessive interest rates are charged, he has our blessings.

One might think that with the vast numbers of dairy farmers going out of business, milk prices would increase, but that is not the case. As a dairyman explained to me, when a dairyman goes out of business, 80–90% of the top producing cows go to a larger producer and the 10–20% go to slaughter. This eventually adds more cows to go through the modern robotic or carousel milking operations.

There are rippling effects of depressed farm prices. Implement dealers in Wisconsin's dairyland also feel the pressure. A dealer whose primary customers are dairy farmers told me that on a nearby road, there were nine dairy farmers just a few years ago and now there are two left. This means that he sells fewer manure spreaders, skid loaders, and other machinery as well as fewer repair parts for all those implements.

This can all be quite depressing, but as in the past, mankind has been able to see the problems and come up with fixes. There is no reason to believe that this won't be done again. As I heard a farm counselor say recently, "Farmers need to understand that when they can't be successful on their small farm, there is another life outside of farming."

In spite of the many challenges, farmers have taken great pride and pleasure in man's oldest occupation. As a farmer I had many challenges. One I haven't mentioned yet is with raising cattle. I was snorted at, pawed at, chased, and kicked at. I was kicked in my legs, kicked in my side, and kicked where the sun never shines.

Since the beginning of time, living off the land had its rewards and challenges. From my observation of farmers, the rewards far out weigh the challenges.

I have great memories of life on the farm and I hope today's farm families are making memories for themselves as well. One of the warm memories of farm life for both me and Ann has been the idea that as we grew up, in spite of hard work and long hours, we seemed to be able to find time to stop and smell the roses along the way with our parents and siblings.

As I was growing up, the parlor in our house, like most houses, had a huge pot-belly stove that would be stuffed full of wood before bedtime in the winter. By morning, there would only be hot coals, but usually enough for my mom or dad, who would be up early, to start a new fire. We kids would hop out of our toasty-warm feather-beds, grab our clothes, dash downstairs, and snuggle up as close to the stove as we could without burning ourselves to get dressed. My older brothers would then hurry to the cow barn where it was quite warm due to the body heat of the 15 cows, all having names, waiting to be milked and fed. There were also about half as many cats gathered around the cat pan patiently waiting for a drink of warm, fresh milk. It couldn't get any fresher than that. The cats, also named, were fat and healthy in spite of never seeing a veterinarian. I am assuming milk and mice were a good diet for fattening cats. Barns were a haven for mice and a perfect breeding place. A good supply of grain was always available. Mice have one thing in common with rabbits; they are good at arithmetic and can multiply fast. This quality kept the cats well supplied with protein.

The kitchen range served many functions. On the right side was a reservoir for warming water for washing dishes, and for whatever we needed hot water. Of course, we needed hot water for our weekly Saturday night baths. Having to carry in and carry out our bath water, a lot less was used for our baths those days. I am told that the tub was filled once for all of us to take a bath in, starting with the oldest kid and working our way down. Ann says she recalls it worked much the same at her house.

The range had an oven in the center. My mom would check the temperature for baking by sticking her hand into the oven. The accuracy of that method of temperature control was questionable but I have fond memories of the smell of fresh loaves of bread and pies baking, excitedly anticipating when I'd be satisfying my taste buds. The loaves of bread were huge and the slices that my mom cut were nearly twice as thick as slices you would get from the grocery store.

The fire box was on the left side with four lids on the top of the stove. The two over the fire on the left would be the hottest and the ideal place to toast our bread. A long fork with two or more tines was stuck into a slice of bread and held directly over the fire after lifting the lid. Hot coals worked the best.

Warming ovens, with two doors, were located about 18 inches above the stove top used to store food to keep it warm. This was often needed by the ladies while waiting for the men folks, coming in late for meals which seemed to be quite common. A teakettle full of water would always be hissing, shooting steam out from its spout. A humidifier was never needed in the kitchen.

The oven-door handle, as well as the handle on the warming oven, were used to hang mittens and gloves that got soaked especially during the snow-fort building and snowball-throwing season. In our house, clothes for drying would be hung on two lines strung from hooks above the summer kitchen door and somehow attached to the top of the stove's warming oven. These clothes lines would often be sagging with heavy, wet shirts and overalls or overhalls as I would often hear them called by some of the folk. Patches on the shirt elbows and on the knees were common. Sometimes the clothes would begin to look as though they were made of patches. It was a bit of a contrast with today when fashion is buying jeans with holes and frays or cutting holes in them intentionally. In those days, we got every penny's worth out of our clothing, sewing patches on patches was quite common. When a bib overall was reduced to near threads, the bib part that didn't see much wear was salvaged for patches after cutting off the pockets.

My final memory of the use of the old cook stove had to do with ironing the clothes. Ironing clothes without electricity, like many other house chores, would be hard to fathom for today's generation. I can still remember my mom and sisters ironing clothes on Saturday evening that we would be wearing to church the next morning. Everything in those days was ironed, even the bed sheets and sometimes the underwear. The ironing board would be set up near the kitchen stove. The three flat irons that we had would be placed on the stove at least an hour ahead of time to get them hot. As ironing got underway, the flat irons would be rotated. When one got cold, it would be placed back on the stove and the cycle continued. Wearing well-ironed clothes was more fashionable but probably out of necessity as we didn't have all the many non-wrinkling fabrics that we have today. The styles were less casual, such as no one would think of wearing jeans to church on Sunday. In fact, no women would wear anything but a dress or skirt.

As I was growing up, I never heard of going out to eat in a restaurant, but with the old kitchen range, my mom and older sisters would make incredible meals, the best meals fit for royalty. I am reminded, however, that memories often become exaggerated.

My siblings and I always looked upon growing up on the farm as one of God's tremendous gifts to us. In spite of the chores, work, inconveniences, dangers, stress, cold, and heat, we considered this way of life to make us among the luckiest people alive.

I am still grateful to have had the learning experiences and skills that farming provided for me and enabled me to pass on to my children. At least they learned at a young age that you don't put a milk bucket under a bull. It appears that they loved their growing up on the farm every bit as much as I did. Their idea for a fun vacation seems to be when they can come home for a weekend and help with any farm chores that needed to be done. While they have gone on to different occupations, they have stayed close to agriculture as much as possible and I think the saying applies to them, "You can take the boy away from the farm but you can never take the farm away from the boy."

Ann and I consider ourselves fortunate to have been raised on a farm and have made a living off the land as well as giving our children the opportunity for such an experience. Although our children are not directly involved in tilling the soil, their employment and hobbies keep them close to the agricultural world.

In the past and present, numerous farm activities have taken place involving the whole family on those occasions when they were all visiting at home. These activities would involve bailing hay, rounding up cattle for vaccinations, weaning, tagging, etc. Weaning calves and running the cows through the runway into the head-gate was a traditional event that took place on the Friday after Thanksgiving.

The first job was to round up the cows into the large yard. This task was the most "funnest" for the older grandkids as they got to ride the ATV's. The next procedure was to separate the calves from the cows. Then the noise of the bawling cows began. Next was to drive the cows, about ten at a time, into a smaller pen and through the runway to the head-gate that automatically locked when the cow poked her head through in an attempt to escape. Now all family members got involved in some way, except for the younger grandkids who were spectators from the back of the pickup, watching with owl-sized eyes, and feeling a bit sorry for the bawling calves.

Pat's and Dean's job was to drive the cows into the runway and prod them into the head-gate. They would carry a small gate in front of themselves to prevent cows from kicking them, thus gaining the nickname, "kickboys." As soon as the cow was caught, Ted's job began. Wearing a long latex glove up to his shoulder, he proceeded to pregnancy-check the cow. He had learned this skill while attending college at Panhandle State University in Goodwell, OK. After feeling around for a calf for 30 seconds or so, he would holler as loud as he could, "pregnant, a nice calf," or "pregnant but late," or "open." By this time, the noise level from the cows bawling, the chute gates clanging, and Ted's yelling as well as family mocking comments and laughter in the background reached higher than the Jack Trice Football Stadium when the Iowa State Cyclones were beating their rival, the Iowa Hawkeyes. Ted would change gloves periodically to try to stay clean, but in spite of this, he had much yucky stuff on his shoulder and running down his shirt, pants and boots. Occasionally, he may have to dodge a yellow waterfall. After Ted would make his pronouncements, the pregnant ones would get their shots. Tim would give a shot in the right side of the neck and I gave one in the left. In the old days, we gave shots in the muscles, but not anymore as we learned that could cause meat damage. I would always consult with Doc Schafer about the latest vaccines and methods being recommended. Tricia's job was to pour a dose of Negevan across their back for parasite control which gave her the nickname, "the Poor Girl." Sometimes instead of just pouring the liquid in a straight line as was recommended by Doc Schafer, Tricia would ignore his recommendation and get creative, making a smiley face or other art designs on the backs of the livestock. At other times, someone may yell that a "bar girl" was needed. That meant Tricia should grab a pipe (bar) and shove it between a cow and two upright posts behind the cow as soon as she entered the runway so that she couldn't back out again in an effort to avoid the head chute. When finished, Tim would open the head-gate. The cows wasted no time in regaining their freedom and running off to look for their calves which, of course, they would not be able to get to any longer because of weaning them. The open cows were herded to a pen for shipping. Tim would also inspect their age, occasionally hollering, "This one is old enough to vote, she won't make the winter." They go to the shipping pen, without having been treated. Someone was given the responsibility of counting the cows; it always seemed this was forgotten until a quarter done so the counts were questionable.

The young grandkids were seeing activity that was foreign to their minds and started asking questions about cows, such as, "When a bull goes through, why didn't Uncle Ted put his arm in that one?" or "When do the steers become cows?" Sometimes, the older kids will be mocking Ted by asking, "What color is the calf?" or "Is it a boy or girl?" or "Hey, Uncle Ted, is your arm staying warm?" or "Hey Ted, does the cow have twins or triplets?"

As noted before, all family members get involved. The older grandkids and in-laws help where needed, such as filling syringes, driving cows into the small yard, helping with prodding cows through the runway which can be challenging, especially with a stubborn cow who has been through the chute before, maybe one that had delivered a breech calf. They seem to sense the traveling through the narrow chute as a threatening experience. However, this is totally safe for the prodders as there are heavy metal bars on both sides of the cow, containing her in a narrow walkway which prevents anyone from being kicked or butted with the head. Every now and then, one of the daughters-in-law would get to be called the Poor Girl or Bar Girl, which may prompt them to suggest they should go to the house and start heating up turkey leftovers for lunch. Remember this was usually taking place the Friday after Thanksgiving.

I recall several years ago, a cow was in labor. I sensed trouble and drove her into the head-gate. Upon examining her, I discovered it was going to be a breech birth as I felt the calf's tail coming first. I called the Veterinarian Associates and explained the problem and was told that Dr. Jen Rediske would be out shortly. Upon arriving—this was her first visit to our farm as she was a recent graduate of Iowa State University and newly hired by the Galena Veterinary Associates—Jen introduced herself. I politely acknowledged her and then addressed her in the manner in which farmers normally talk to their vets. I said, "Doc, I hope you have a stronger arm than I do." Her response was, "I'll get the calf delivered." Dealing with breech calves, Dr. Jen had a unique way of going into the cow with both arms pushing the calf forward with one hand and straightening the legs with the other. Within about ten minutes, the approximately 80 lb. bull-calf was born and within another 15 minutes, he was having the first meal of his life.

So this ends my stories, some sad, some funny, stories of remembered hard work, some just fond memories but this much I know, writing this book was much like farming, i.e. challenging, fun, a learning experience, hard work, but for me, a true labor of love.

Glossary

BARROW PIG: A male pig (castrated) rendered incapable of reproducing.

BROKEN MOUTH SHEEP: Sheep with the premature loss of incisor teeth leading to early involuntary culling because affected sheep are unable to bite off short and/or rough pasture leading to malnutrition, poor production, and weight loss.

CANDLING EGGS: Using a bright light behind the egg to check if any impurities are present or if an embryo is present. It is so named because originally a candle was used as the source of light.

CLUCKY HEN: A clucky or broody hen is one that faithfully wants to sit on her eggs until they hatch, which is 21 days.

COCKERALS: A young male chicken.

EWE: A female adult sheep. A word that is rarely used or understood as so few people are engaged in raising sheep. It is a more popular word used in crossword puzzles.

FLAIL: A long-handled tool used to separate grain from the stock.

GERM: The embryo of a plant seed.

GILT: A female hog under the age of one year. When she gives birth, she becomes a sow.

GRIT: Processed or ground oyster shells that aid in digesting food and provide for laying eggs with harder shells. Small pebbles can perform the same function which is why you might see chickens appearing to be eating gravel.

GIRDLING TREES: Sawing a two-inch cut into the tree trunk all around it to stop the flow of sap. It is recommended to make a double cut about ten inches apart for hard to kill trees such as mulberries, black and honey locust, boxelder, and maple trees.

HAGGERT: Meat from a sheep between the age of one and two years old.

LAMB MEAT: Meat from a lamb under one year old.

MASH: Ground and mixed feed for chicks.

MOCHILA: A square leather covering that slings over the saddle with openings for the saddle horn and cantle with four pockets. It was more familiarly used by the Pony Express riders to carry the mail. Two pockets were locked and not to be opened until reaching San Francisco. Two were to carry deliveries to be made along the way.

MOLTING: The natural process of chickens and birds shedding old feathers.

MUTTON: Meat from sheep over two years old.

PULLETS: Young female chickens under the age of one year.

ROOSTER SPUR: A sharp, horn-like growth on the back of a rooster's leg which he uses for defense.

SERENDIPITY: The unintentional good or useful discovery, such as in health remedies or crop aides.

TILLERS: Corn suckers that shoot out from the bottom of corn stalks. They are an undesirable trait that take nutrition from the plant.

Acknowledgments

My wife, Ann, who contributed the most —the typing, especially grammar and spelling corrections as they have been a lifetime mystery for me.
My children, my co-authors:

Ted

Tim

Pat

Tricia

Dean

whose brains I picked for invaluable suggestions and information on all chapters.
My 95 year old brother, Richard, with his impeccable memory of "the good old days" shared with me such things as picking corn by hand.
Mike Gibson, Loras College Historian, who helped me obtain articles and information.
The Stonefield Museum at Cassville, WI for allowing me to take pictures.
Cal Schaffer, D.V.M., for contributing much to the chapter, "Livestock Sickness and Treatments" and allowing me to use his 1889 fragile book, *Gleason's Veterinary Hand Book and System of Horse Training*.
Dan Brown and Anne Brown, of Liberal, KS, who shared stories of raising wheat and pasturing feeder cattle and about rattlesnakes.
KelliJo Brown, daughter-in-law of Dan and Anne Brown, who provided grain harvest pictures (kj's sure shots by KellyJo Brown) on the cover and in the center with the colored pictures.
Linda Digman, for lending me her 1879 faded book, The History of Jo Daviess County Illinois.

122 Changes In the Good Life

All the folks at the Lawfer Farm, who gave Ann and me a lengthy tour of their newly installed robotic milking operation and explaining its high-tech functions.

Dan, David, and John Carroll, especially David, who gave up time to talk about their hog and grain operations in the U.S. and their farming adventure in Brazil.

Tom & Judy Schuldt and their son & daughter-in-law, Jeff and Heidi Schuldt, who allowed taking pictures of picking corn.

Dr. Robert Cropp, Dairy Specialist, who contributed data about dairy farming.

Karen Meyer, Menominee's J & M Tap for some Menominee history.

Clinton White, for sharing his experiences of hauling the farmers' milk to the Menominee Creamery.

Phil Gent of the Jackson Co. Museum for allowing me to take pictures and for his wealth of information.

Andrea Bednar at River Lights Publishing for her assist in getting this book published as I wanted and especially for her patience with me.

In 1937, Bert Tranel was born on the 127 acre family farm four miles from East Dubuque, IL, the youngest of 13 children. His father passed away four and a half months after his birth.

After first grade at a one room country school, he attended St. Mary's Catholic School in East Dubuque. His high school education was at Loras Academy, Catholic College in Dubuque, IA.

After one year at Loras College, he began to pursue his life-long dream of farming.

In 1963, he married Ann Berlage. That marriage produced five children and nine grandchildren.

He has continued his farming career, raising corn, soybeans, oats, and hay. He also maintains a herd of Black Angus Cattle.

This is Bert's second book. His first book was, *"Lucky 13," Farm Family of Faith*.